Student Assessments with Solution Keys and Scoring Guides

The assessments for this text, **MATH** *Connections* Year I, reflect the intent of the NCTM Standards on Assessment in several ways.

• The assessments allow students to use a variety of formats to demonstrate their knowledge of mathematics concepts and skills. Students are consistently required to analyze problems, determine solutions and write justifications for their solutions. Some quizzes have performance assessment items that may be assigned to a group or to an individual.

• The assessments are tied to specific mathematics objectives. Each assessment item is directly linked to the Learning Outcomes for each section of the chapter. The assessment items on the chapter test may not match every learning outcome for a chapter.

• The assesssments focus on communication, reasoning, problem solving and connections as well as skill development. They reflect the concepts, content and processes of mathematics.

• The Solution Key and Scoring Guide for each quiz and chapter test contains solutions to all questions as well as *suggestions* for scoring based on a 100 point scale. A rubric has been developed for open-ended questions, which details the elements of a rich response.

Two assessment forms, Form A and Form B, were developed for each quiz and chapter test. The two forms are designed to be of comparable difficulty and provide you with an alternative assessment for each quiz and chapter test.

— Donald Hastings
Assessment Specialist
Stratford Public Schools (retired)

Assessments Form A

Assessments Form A & B — Are designed to allow students to use a variety of formats to demonstrate their knowledge of mathematics concepts and skills. Assessments are tied to specific mathematics objectives and are linked to the Learning Outcomes for each section of the chapter. Two Assessment forms, Form A and Form B, were developed for each quiz and chapter test. The Solution Key and Scoring Guide for each quiz and chapter test contains solutions to all questions as well as suggestions for scoring based on a 100 point scale.

Legend:

The header to each Assessment is explained as follows:

Upper left corner: **MATH** *Connections* I refers to Year 1.

Upper right corner: Quiz 1.1–1.2 (A) refers to a quiz for Chapter 1, Sections 1.1–1.2, Form A. The reference sk: Quiz: 1.1–1.2 (A) refers to the Solution Key and Scoring Guide for the Quiz for Chapter 1, Sections 1.1–1.2, Form A.

This project was supported, in part, by the
National Science Foundation
Opinions expressed are those of the authors
and not necessarily those of the Foundation.

ISBN 1-891629-56-5
ISBN 1-58591-030-9
Published by IT'S ABOUT TIME, Inc. © 2000 MATHconx, LLC

MATH *Connections* I
Using Lines and Equations Quiz
Sections 5.1 - 5.3 (A)

Name _____ Date _____

1. Tommy asked Jose to guess the number between 1 and 12 of which Tommy was thinking. Jose asked several questions and finally determined it was 7. Make a tree which shows which questions Jose asked to discover that 7 was the answer.

2. Sarah says she will treat you to a hamburg and milkshake if you can guess the number she is thinking between 0 and 200 in six or less guesses.
   What should your first question be if you wish to rule out half (or more) of the possible numbers that she could have chosen?

   _____

   _____

   _____

3. Two ice cream parlors in New Morris decided to have a special to attract birthday party business. Aunt Mary's Ice Cream Parlor charged $2.50 for each sundae and $4.00 for a rental fee. To compete Pedro's Screaming Ice Creams charged $2.00 for each ice cream sundae and $6.00 for a rental fee.

   a. If you and one friend celebrate your birthday, which ice cream parlor would you choose if you want to save money? Demonstrate why you chose your answer.

b. If you and seven friends were going to celebrate your birthday, which ice cream parlor would you choose if you want to save money? Demonstrate why you chose your answer.

c. Write an equation in the form of y = mx + b to determine the cost of a birthday party in each ice cream parlor. Use C to represent the cost and S to represent the number of sundaes sold.

Aunt Mary's _____

Pedro's _____

d. Complete the table below to show the comparative cost of a birthday celebration in each store:

Number of Sundaes	Aunt Mary's	Pedro's
1		
2		
3		
4		
5		
6		
7		
8		
9		
10		

e. How many sundaes could you order so that the cost of the party was the same in each parlor? _____

What would that cost be? _____

f. Graph each equation on your calculator. Copy the graphs onto a piece of graph paper.

g. Explain how you can determine or estimate the intersection point of the two graphs using your calculator.

_____

_____

_____

_____

_____

_____

4. For the following pairs of numbers,
   • find the number which is halfway between them
   • for (a) and (b), on a number line show the two given numbers and the number halfway between.

   a. 5 and 8   _____   ⟵———————————————⟶

   b. -7 and 13   _____   ⟵———————————————⟶

   c. 4.26 and 8.58   _____

   d. -1.374 and 3   _____

5. Find the fourth root of 9.

SOLUTION KEY AND SCORING GUIDE

MATH *Connections* I
Using Lines and Equations Quiz
Sections 5.1 - 5.3 (A)

1. Tommy asked Jose to guess the number between 1 and 12 of which Tommy was thinking. Jose asked several questions and finally determined it was 7. Make a tree which shows which questions Jose asked to discover that 7 was the answer. *(6)*

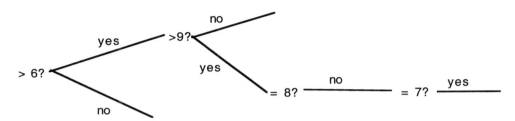

Note: Jose could ask "= 7?" before "= 8?" and saved one step.

2. Sarah says she will treat you to a hamburg and milkshake if you can guess the number she is thinking between 0 and 200 in six or less guesses.
What should your first question be if you wish to rule out half (or more) of the possible numbers that she could have chosen? *(6)*

   Is the number more than (or less than) or (equal to) 100?

3. Two ice cream parlors in New Morris decided to have a special to attract birthday party business. Aunt Mary's Ice Cream Parlor charged $2.50 for each sundae and $4.00 for a rental fee. To compete Pedro's Screaming Ice Creams charged $2.00 for each ice cream sundae and $6.00 for a rental fee.

   a. If you and one friend celebrate your birthday, which ice cream parlor would you choose if you want to save money? Demonstrate why you chose your answer.
   (10)
   Aunt Mary's: 2.5 • 2 + 4 = 9

   Pedro's: 2 • 2 + 6 = 10   Choose Aunt Mary's.

b. If you and seven friends were going to celebrate your birthday, which ice cream parlor would you choose if you want to save money? Demonstrate why you chose your answer. *(10)*

      Aunt Mary's: 8 • 2.5 + 4 = 24
      Pedro's: 8 • 2 + 6 = 22
      You would choose Pedro's.

c. Write an equation in the form of y = mx + b to determine the cost of a birthday party in each ice cream parlor. Use C for cost and S for number of sundaes sold.
(10)
Aunt Mary's *C = 2.5 S + 4*

Pedro's *C = 2 S + 6*

d. Complete the table below to show the comparative cost of a birthday celebration in each store: *(10)*

Number of sundaes	Aunt Mary's	Pedro's
1	6.5	8
2	9.0	10
3	11.5	12
4	14	14
5	16.5	16
6	19	18
7	21.5	20
8	24	22
9	26.5	24
10	29	26

e. How many sundaes could you order so that the cost of the party was the same in each parlor? _4_ *(10)*

What would the cost be? _$14_

f. Graph each equation on your calculator. Copy the graphs onto a piece of graph paper.
(6)

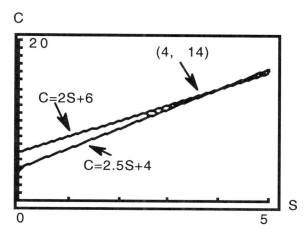

g. Explain how you can determine or estimate the intersection point of the two graphs using the graphing calculator. *(6)*

*Use ZOOM and TRACE functions to find an estimate of the point where the two lines intersect. **or***

Use the CALC and INTERSECT functions to find an exact point of intersection.

4. For the following pairs of numbers, find the number which is halfway between them. For (a) and (b), on a number line show the two given numbers and the number halfway between.
(16)

a. 5 and 8

_6.5_

b. -7 and 13

_3_

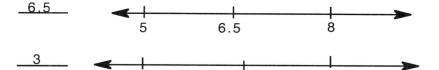

c. 4.26 and 8.58

_6.42_

d. -1.374 and 3

_.813_

5. Find the fourth root of 9.   *(10)*
Full credit should reflect an argument similar to the one below.
$1^4 = 1$, while $2^4 = 16$. Therefore the fourth root of 9 is between 1 and 2.
$(1.5)^4 = 5.0625$ which tells us that the fourth root of 9 is closer to 2.
$(1.75)^4 = 9.38$ which tells us that the answer is less than 1.75.
The teacher will determine how close to the actual answer is expected. The actual answer is 1.73.

MATH *Connections* I
Using Lines and Equations Quiz
Sections 5.4 - 5.5 (A)

Name _____ Date_____

1. For the following pairs of linear equations,
   - graph both equations on your calculator and
   - by using the CALC function determine the coordinates of their intersection.

   a. $Y_1 = -0.2x + 4$

       $Y_2 = 1.3x - 1$

   b. $Y_1 = 2x - 8$

       $Y_2 = 3x + 6$

   _____   _____

   Explain the steps you used with the calculator to determine the intersection.

2. Aunt Mary's Ice Cream Parlor advertises the store's birthday special at $2.50 for each sundae and a $4.00 rental fee.

   a. Write an equation to determine the total cost of a party based on the number of sundaes to be served.
      - Use C to represent the total cost of the party, and
      - S to represent the number of sundaes . _____

   Is C or S the independent variable? _____ Explain.

   _____

   _____

   b. How many sundaes did Aunt Mary prepare for each of the birthday parties listed below? (Show all work).

   i. The total cost of Tim's party was $54.00.

   ii. The total cost of Shana's party was $36.50.

iii. The total cost of Rani's party was $44.00.

c. On your calculator, graph the equation you wrote for your answer to question 2(a) Copy the graph on to a piece of graph paper.

Determine the C (or y) intercept. _____

Determine the S (or x) intercept. _____

3. For overnight express Speedy Delivery Service charges $3.00 for handling any package plus $0.60 for each ounce the package weighs (to the nearest ounce). Using C for cost and W for the weight of each package in ounces, write an equation for determining the cost for overnight packages. _____

Show all work used to determine the weight of each package given that the total cost is:

a. $9.00   b. $111.00   c. $40.20

4. Solve each of the following equations for the variable that is designated. Show all work.

a. $5x + 2y = 12$ for y.   b. $200h + 600t = 24000$ for h

5. On January 1, Maria has to make a job decision. Maria enjoys her job with Martin and Smith Attorneys and will make $32,000 next year.   Martin and Smith can guarantee a 5% raise each year for the next five years.

The Five Star law firm promises her an immediate increase of $2000 for joining their firm and a 2.5% increase in salary for the next five years.

Maria plans to sign a five year agreement, as she is enrolled in a law degree program that will take her five years to complete.   Maria asks you for advice on which agreement she should sign.

To assist in your decision, you complete the table below.

Year	Martin and Smith	Five Star
1	32000	34000
2	_____	_____
3	_____	_____
4	_____	_____
5	_____	_____

a. • Place the three columns in DATA lists L1, L2 and L3.

   • Make a scatter plot of each salary by using L1 and L2 for one plot and L1 and L3 for the other plot. Be sure to use different symbols for the points on each plot.

   • Make the window ranges as follows:

       Xmin = 0   Ymin = 30000
       Xmax = 7   Ymax = 42000
       Xscl = 1   Yscl = 2000

 • Copy the scatter plots on graph paper and connect the points for each plot.

   b. Using the data in your table and any other reasons you feel appropriate, advise Maria on which offer she should take.

_____

_____

_____

_____

SOLUTION KEY AND SCORING GUIDE

MATH *Connections* I
Using Lines and Equations Quiz
Sections 5.4 - 5.5 (A)

1. For the following pairs of linear equation,
   - graph both equations on your calcullator _and_
   - by using the CALC function determine the coordinates of their intersection.
   (5 each)

   a. $Y_1 = -0.2x + 4$
   $\ \ Y_2 = 1.3x - 1$

   b. $Y_1 = 2x - 8$
   $\ \ Y_2 = 3x + 6$

   __(3.333..., 3.333...)__

   _(-14, -36)_

   Explain the steps you used with the calculator to find the intersection. **(6)**

   Assuming the calculator has the STANDARD window display, in part (a), the CALC menu is selected and the INTERSECTION option is chosen. The cursor is moved as close to the intersection as possible and ENTER is selected 3 times. The intersection is then displayed. In part (b), the window has to be adjusted to include the intersection. Then the procedure for finding the intersection is the same as in part (a).

2. Aunt Mary's Ice Cream Parlor advertises the store's birthday special at $2.50 for each sundae and a $4.00 rental fee.

   a. Write an equation to determine the total cost of a party based on the number of sundaes to be served. Use C to represent the total cost of the party, and S to represent the number of sundaes . $\underline{C = 2.5\,S + 4}$ **(5)**

      Is C or S the independent variable? __S__ Explain. **(5)**

      The cost of the party (C) is determined by the number of sundaes (S) that are purchased. This makes S the independent variable or C dependent.

   b. How many sundaes did Aunt Mary prepare for each of the birthday parties listed below? (Show all work). **(5 each)**

   i. The total cost of Tim's party was $54.00.
      S = 20

   ii. The total cost of Shana's party was $36.50.

   S = 13

1

iii. The total cost of Rani's party was $44.00.

$S = 16$

c. On your calculator, graph the equation you wrote for your answer to question 2(a). Copy the graph on to a piece of graph paper. *(3)*

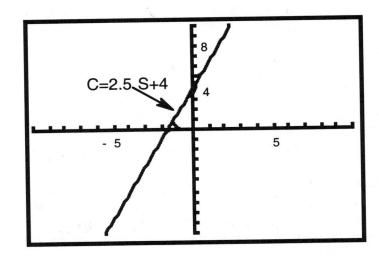

Determine the C (or y) intercept. _(0, 4)_____ *(3)*

Determine the S (or x) intercept. ___(-1.6, 0)_ *(3)*

3. For overnight express Speedy Delivery Service charges $3.00 for handling any package plus $0.60 for each ounce the package weighs (to the nearest ounce). Using C for cost and W for the weight of each package in ounces, write an equation for determining the cost for overnight packages. __C = .60 W + 3.00__ *(4)*

   Show all the work used to determine the weight of each package given that the total cost is: *(4 each)*

   a. $9.00   b. $111.00   c. $40.20

   _W = 10_   _W = 180_   _W = 62_

4. Solve each of the following equations for the variable that is designated. Show all work. *(4 each)*

   a. 5x + 2y = 12 for y.   b. 200h + 600t = 24000 for h

   $y = -2.5x + 6$ $h = -3t + 120$

5. On January 1, Maria has to make a job decision. Maria enjoys her job with Martin and Smith Attorneys and will make $32,000 this next year. Martin and Smith can guarantee a 5% raise each year for the next five years.

The Five Star law firm promises her an immediate increase of $2000 for joining their firm and a 2.5% increase in salary for the next five years.

Maria plans to sign a five year agreement as she is enrolled in a law degree program that will take her five years to complete. Maria asks you for advice on which agreement she should sign.

To assist in your decision you complete the table below. *(10)*

Year	Martin and Smith	Five Star
1	32000	34000
2	33600	34850
3	35280	35721
4	37044	36614
5	38896	37529

a. • Place the three columns in DATA lists L1, L2 and L3.

   • Make scatter plot of each salary by using L1 and L2 for one plot and L1 and L3 for the other plot. Be sure to use different symbols for the points on each plot.

   • Make the window ranges as follows:

   Xmin = 0   Ymin = 30000

   Xmax = 7   Ymax = 42000

   Xscl = 1   Yscl = 2000

• Copy the scatter plots on graph paper and connect the points for each plot. **(4)**

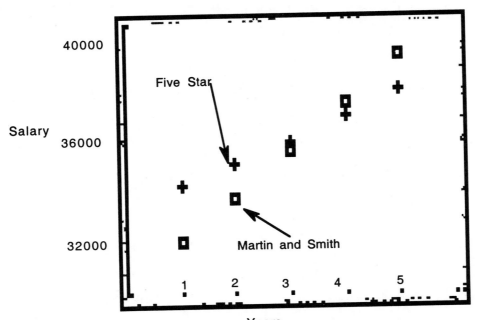

b. Using the data from your table and any other reasons you feel appropriate, advise Maria on which offer she should take. **(12 points)**

A rich answer should contain some of the following arguments:

1. The total amount of money she receives from Martin and Smith is $176,820 while the total amount of money she receives from Five Star is $178,714.

2. She is ahead for two years while working for Five Star, but is ahead three years for Martin and Smith.

3. The final decision could easily go either way. The student will receive credit for the best argument.

MATH *Connections* **I**
Using Lines and Equations Quiz
Sections 5.6 - 5.7 (A)

Name _____ Date _____

1. Use your calculator to solve each of the following equations:

   a. $3(x - 5) = 14$ b. $220x + 82 = 473$

   c. $3200\,(1.03)^n = 42000$

2. Pedro owns Pedro's Screaming Ice Cream and charges $2.00 for each sundae and a $6.00 rental fee for birthday parties. At the end of a very busy Saturday Pedro had $680 in the birthday party account. Pedro knows there were 20 different birthday parties. He would like to determine how many people attended the parties. Assist Pedro by answering the following questions.

   a. How much money was collected for just the rental fees? _____

   b. Write an equation that shows that for the 20 birthday parties, $680 was collected which includes rental fees and the $2.00 that was charged for each sundae sold during a party. _____

   c. Use your calculator to solve this equation to find the total number of people who attended the parties. _____

1

3. Find the solution for each of the following pairs of equations. Show all work.

a. $y = 4x + 10$
    $y = 3x - 6$

b. $S = -2R - 10$
    $S = 9R + 12$

4. Aunt Mary charges $2.50 for each sundae and a $4.00 rental fee, while Pedro charges $2.00 for each sundae and a $6.00 rental fee.
    a. Write the equation that describes the money each received for a birthday party. In each equation use C to represent the cost and S to represent the number of sundaes sold.

    Aunt Mary _____   Pedro _____

    b. Solve the equations simultaneously for S.

    c. What does your value of S represent?

5. The Hard Rock Music Company makes compact discs and audio tapes. The Company makes $6.00 profit on each compact disk(CD) and $4.00 on each tape.

    a. Write an algebraic equation for the profit (P) in terms of the compact disks (C) and tapes (T).

    _____

    b. The company can produce a total of 500 tapes and CD's. Write an equation that reflects
        this. _____
    Graph this equation on a piece of graph paper.

    c. It costs $10.00 to produce each CD and $8.00 to produce each tape. The company can pay only $4800.00 for production costs. Write an equation that reflects this. _____ Graph this equation on the same set of axes that you graphed 5(b).

d. From your graph find the coordinates of the point of intersection point of the two lines: _____

e. Use your profit equation from part (a). to determine the profit at the point you found in part (d). Show your work.

SOLUTION KEY AND SCORING GUIDE

MATH *Connections* I
Using Lines and Equations Quiz
Sections 5.6 - 5.7 (A)

1. Use your calculator to solve each of the following equations: *(7 each)*

   a. $3(x - 5) = 14$ b. $220x + 82 = 473$

   $x = 9.66...$ *$x = 1.7772727...$*

   c. $3200 (1.03)^n = 42000$

   n = approximately 87

2. Pedro owns Pedro's Screaming Ice Cream and charges $2.00 for each sundae and a $6.00 rental fee for birthday parties. At the end of a very busy Saturday Pedro had $680 in the birthday party account. Pedro knows there were 20 different birthday parties. He would like to determine how many people attended the parties. Assist Pedro by answering the following questions. *(7 each)*

   a. How much money was collected for just the rental fees?

      $120.00

   b. Write an equation that shows that for the 20 birthday parties, $680 was collected which includes rental fees and the $2.00 that was charged for each sundae sold during a party a party. *$680=$2S+$120*

   c. Use your calculator to solve this equation to find the total number of people who attended the parties. *$S = 280$*

3. Find the solution for each of the following pairs of equations: (Show all work.)
    (6 each)
    a. $y = 4x + 10$ b. $S = -2R - 10$
          $y = 3x - 6$ $S = 9R + 12$
          4x + 10 = 3x - 6   *-2R - 10 = 9R + 12*
          x = -16   *-11R = 22*
          y = -54   *R = -2 ; S = -6*

4. Aunt Mary charges $2.50 for each sundae and a $4.00 rental fee while Pedro
   charges $2.00 for each sundae and a $6.00 rental fee.
   a. Write the equation that describes the money each received for a birthday party.
      In each equation use C to represent the cost and S to represent the number of
      sundaes sold. **(7 each)**

         Aunt Mary *C = $2.5 S + $4*   Pedro *C = $2 S + $6*

   b. Solve the equations simultaneously for S.

            2.5 S + 4 = 2 S + 6
                 .5 S = 2
                   S = 4

   c. What does your value of S represent?

         S represents the number of sundaes that would be purchased at
         either Aunt Mary's or Pedro's for which the cost is the same.

5. The Hard Rock Music Company makes compact discs and audio tapes. The
   Company makes $6.00 profit on each compact disk(CD) and $4.00
   on each tape.   **(5 each)**

   a. Write an algebraic equation for the profit (P) in terms of the compact disks (C) and
       tapes (T). *P = 6 C + 4 T*

   b. The company can produce a total of 500 tapes and CD's. Write an equation that
      reflects this. *C + T = 500*
      Graph this equation.

   c. It costs $10.00 to produce each CD and $8.00 to produce each tape.
      The company can pay only $4800.00 for production costs. Write an
      equation that reflects this.   *10 C + 8 T = 4800*
      Graph this equation on the same set of axes that you graphed 5(b).

d. From your graph find the coordinates of the point of intersection point of the two
   lines: *(100, 400)*

e. Use your profit equation from part (a). to determine the profit at the
   point you found in part (d). Show your work.

   P = 6 • 400 + 4 •100
   P = 2800

Graph for 5(b) and (c)

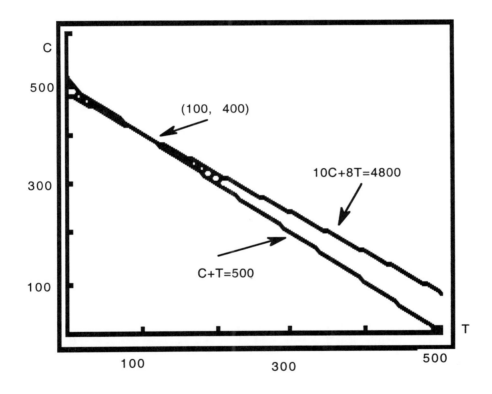

MATH *Connections* I
Using Lines and Equations Test
Chapter 5 (A)

Name _____ Date _____

1. Mary had to leave the Park High School basketball game at the end of the third period with the score tied 50 to 50. Next day she asked Rafael what the final score was. Rafael told her that Park had won and had scored no more than 80 points. Rafael asked Mary to guess the Park score.

   a. What should Mary's first question be if she wishes to rule out half (or more) of the possible scores?

   b. How many guesses do you think Mary would need? _____ Make a tree to justify your answer.

   c. On a number line show the number that is halfway between 50 and 80.

2. Solve each of the following equations for the variable that is designated. Show all work.

   a. $5x - 3y = 15$ for y

   b. $25s + 45t = 600$ for t

3. Geoffrey is from Australia where the metric measurement system is used. He told the Connections class at Jurassic Park High School that an easy way to approximate the Fahrenheit temperature is to double the Celsius temperature and add 30° to the product.

a. Write an equation for the method that Geoffrey uses to approximate a Fahrenheit temperature when you know a Celsius temperature. Use C for Celsius and F for Fahrenheit.   _____

b. Which variable, C or F, is the independent variable? _____

c. The exact conversion equation is $F = 1.8 \cdot C + 32$. Using this equation determine the

C and F intercepts. Show all your work.

<u>C intercept</u>   <u>F intercept</u>

d. Use your calculator to graph the equation you found in part (a) and the equation given in part (c), $F = 1.8 \cdot C + 32$.

e. Use your calculator to find the Celsius temperature at which both equations give the same Fahrenheit temperature. _____

f. Demonstrate that your answer produces the same result in each equation.

g. Using the exact conversion equation, find the Celsius temperature for the given Fahrenheit temperatures:
   Show all work.
   i. $F = 80°$ ii. $F = -15°$

4. The wind chill factor during the winter is a serious consideration for people who work outdoors. The local meteorologist gave this partial list of windchill factors for a temperature of 25° F:

Wind Speed(mph)	Windchill (°F)
5	21
10	10
15	2

a. Load this data in L1 and L2 in the Stat menu.

b. Make a scatter plot on the calculator and transfer it to your graph paper.

c. Use STAT/CALC/LinReg(ax+b) menus to find the slope and y-intercept of the line of best fit. Write the equation of the line in y = mx + b form: _____

d. Use this line or equation to predict what the temperature would feel like at 40 mph: _____

e. If the temperature feels like -10° F, use the line or equation to predict the wind speed: _____

f. Maria found a complete table of windchill factors for 25°F in the 1994 World Almanac. It is listed below:

Wind Speed (mph)	Windchill (°F)
5	21
10	10
15	2
20	-3
25	-7
30	-10
35	-12
40	-13
45	-14

Load these into L1 and L2 and now use STAT/CALC/LinReg(ax+b) to determine the slope and y-intercept of the line of best fit.

g. The equation of this line is: _____

h. Explain the difference in the two prediction lines. What did you learn about the windchill factor?

_____

_____

_____

_____

_____

5. A Washington State fruit grower has 600 acres of land to grow apples and grapes.

   a. Write an equation that describes the number of acres of apples (A) and grapes (G). _____

   - Find the G-intercept of the graph of this equation. _____

   - Find the A-intercept of the graph of this equation. _____

   - Graph this equation on graph paper supplied by your teacher.

   The profit per acre for apples is $40 and the profit per acre of grapes is $30.

   b. Write an equation for profit (P) made on the apples (A) and grapes (G).

   _____

   The total labor hours available during the harvesting is 8000 hours.   Each acre of apples uses 10 hours of labor, while each acre of grapes uses 20 hours of labor.

   c. Write an equation that describes the amount of labor used to harvest the fruit.

   _____

   - Find the G-intercept for the graph of this equation. _____

   - Find the A-intercept for the graph of this equation. _____

   - Graph this equation on the same graph.

   d. Determine the point where the graphs of the equations of parts (a) and (c) meet.

   _____

   e. Determine the profit at this point. Show your work.

Extra Credit: Using Geoffrey's idea in question 2, how could you approximate the Celsius temperature when you know the Fahrenheit temperature? What easy way do you suggest?

SOLUTION KEY AND SCORING GUIDE

MATH *Connections* I
Using Lines and Equations Test
Chapter 5 (A)

1. Mary had to leave the Park High School basketball game at the end of the third period with the score tied 50 to 50. Next day she asked Rafael what the final score was. Rafael told her that Park had won and had scored no more than 80 points. Rafael asked Mary to guess the Park score.

    a. What should Mary's first question be if she wishes to rule out half

        (or more) of the possible scores?   *(3)*

       She should ask "Was Park's score more (or less) than 65 (or 64)".

    b. How many guesses do you think Mary would need? _____   *(2)*
    Make a tree to justify your answer.   *(3)*

       The answer should be 5. The tree could take several forms, depending on the way the students have Mary asking the questions.

    c. On a number line show the number that is halfway between 50
       and 80.   *(2)*

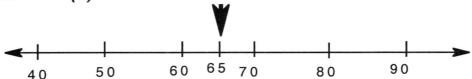

2. Solve each of the following equations for the variable that is designated. Show all work.   *(3 each)*

    a. $5x - 3y = 15$ for y   b. $25s + 45t = 600$ for t

       $y = \dfrac{5}{3}x - 5$ $t = \dfrac{40}{3} - \dfrac{5}{9}s$

2. Geoffrey is from Australia where the metric measurement system is used. He told the Connections class at Jurassic Park High School that an easy way to approximate the Fahrenheit temperature is to double the Celsius temperature and add 30° to the product.

a. Write an equation for the method that Geoffrey uses to approximate a Fahrenheit temperature when you know a Celsius temperature. Use C for Celsius and F for Fahrenheit. *(4)*

$\underline{F = 2C + 30}$

b. Which variable, C or F, is the independent variable? $\underline{C}$ *(2)*

c. The exact conversion equation is F = 1.8 C + 32. Using this equation determine the C and F intercepts. Show all your work.   *(3 each)*

C intercept	F intercept
$0 = 1.8 \bullet C + 32$	$F = 1.8 \bullet 0 + 32$
$-32 = 1.8 \bullet C$	$F = 32$
$-17.77... = C$	$(0, 32)$
$(-17.8, 0)$	

d. Use your calculator to graph the equation you found in part a. and the equation given in part c.

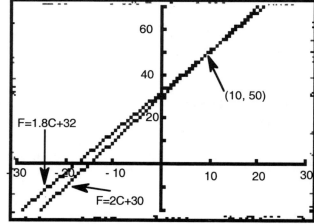

e. Use your calculator to find the Celsius temperature at which both equations give the same Fahrenheit temperature. $\underline{C = 10}$ *(4)*

f. Demonstrate that your answer produces the same result in each equation. *(3)*

$F = 2 \bullet 10 + 30$
$F = 50$

$F = 1.8 \bullet 10 + 32$
$F = 50$

g. Using the exact conversion equation, find the Celsius temperature for the given Fahrenheit temperatures: Show all work. *(2 each)*
   i. $F = 80°$ ii. $F = -15°$

$80 = 1.8 \bullet C + 32$
$48 = 1.8 \bullet C$
$26.7 = C$

$-15 = 1.8 \bullet C + 32$
$-47 = 1.8 \bullet C$
    $-26.1 = C$

3. The wind chill factor during the winter is a serious consideration for people who work outdoors. The local meteorologist gave this partial list of windchill factors for a temperature of 25° F:

Wind Speed (mph)	Windchill (°F)
5	21
10	10
15	2

a. Load this data in L1 and L2 in the Stat menu.

b. Make a scatter plot on the calculator and transfer it to your graph paper. *(4)*

c. Use STAT/CALC/LinReg(ax+b) menus to find the slope and y-intercept of the line of best fit. Write the equation of the line in $y = mx + b$ form: *y = - 1.9 x + 30* *(4)*

d. Use this line or equation to predict what the temperature would feel like if the wind is 40 mph: *- 46°*   *(4)*

e. If the temperature feels like -10° F, use the line or equation to predict the wind speed: *21 mph*   *(4)*

f. Maria found a complete table of windchill factors for 25°F in the 1994 World

Almanac. It is listed below:

Wind Speed (mph)	Windchill (°F)
5	21
10	10
15	2
20	-3
25	-7
30	-10
35	-12
40	-13
45	-14

Load these into L1 and L2 and now use STAT/CALC/LinReg(ax+b) to determine the

slope and y-intercept of the line of best fit.

g. The equation of this line is: _y -0.8 x +17_   **(4)**

h. Explain the difference in the two prediction lines. What did you learn about the

windchill factor?   **(6)**

Using more data gives a more accurate prediction line. The original three points
were almost linear and did not reflect an accurate picture of the behavior of the
windchill factor. The windchill factor changes very little above 25° mph.

4. A Washington State fruit grower has 600 acres of land to grow apples and grapes.

a. Write an equation that describes the number of acres of apples (A) and grapes

(G). _A + G = 600_   **(4)**

Find the G-intercept of the graph of this equation. *(0,600)*   **(2)**

Find the A-intercept of the graph of this equation. *(600,0)*   **(2)**

Graph this equation on the graph paper. **(4)**

The profit per acre for apples is $40 and the profit per acre of grapes is $30.

b. Write an equation for profit (P) made on the apples (A) and grapes (G).*(4)*

$P = 40\,A + 30\,G$

The total labor hours available during the harvesting is 8000 hours. Each acre of apples uses 10 hours of labor while each acre of grapes uses 20 hours of labor.

c. Write an equation that describes the amount of labor used to harvest the fruit. *(4)*

$10\,A + 20\,G = 8000$

Find the G-intercept for the graph of this equation. *(0, 400)*   *(2)*

Find the A-intercept for the graph of this equation. *(800, 0)*   *(2)*

Graph this equation on the same graph. *(4)*

d. Determine the point where the graphs of the equations of parts (a) and (c) meet.

(4)

$A = 400;\ G = 200$

e. Determine the profit at this point. Show your work.   *(3)*

$P = 40 \bullet 400 + 30 \bullet 200$

$P = 22000$

Graph:

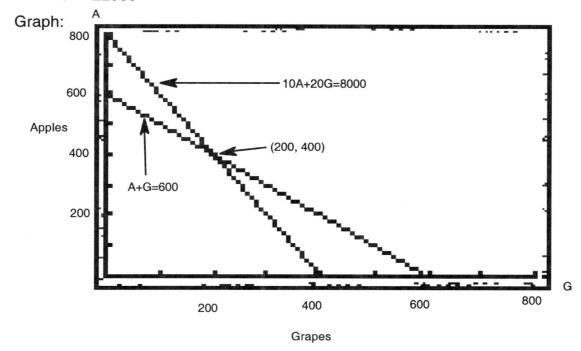

Extra Credit: Using Geoffrey's idea in question 2, how could you approximate the Celsius temperature when you know the Fahrenheit temperature? What easy way do you suggest? *(5)*

The easiest way would be to subtract 30 from the Fahrenheit temperature and divide the result in half. That would give you an approximate Celsius temperature.

MATH *Connections* I
How Functions Function Quiz
Sections 6.1 - 6.2 (A)

Name _____ Date _____

1. Tammy borrowed $7000.00 from her parents to buy a car. Her parents agreed that
    she can pay $500.00 on January 1 and $500.00 on July 1 each year to repay the
    loan.

a. The first few payments are in the chart below. Complete steps for
    payments 3, 4 and 5.

Payments		Unpaid Balance $7000.00
First Payment: Jan. 1	$500.00	$6500.00
Second Payment: July 1	$500.00	$6000.00
Third Payment: Jan. 1		
Fourth Payment: July 1		
Fifth Payment: Jan. 1		

b. Write a function for the unpaid balance (B) of the loan in terms of the
    number of loan payments (n).

    _____

c. What is the domain of this function? _____

d. What is the range of this function? _____

e. What will be the unpaid balance after the 10th payment? _____

f.   Draw a graph that represents the unpaid balance and the first 5 payments on the
     loan.

2. Tammy offers to pay some interest to her parents on the money she borrowed. She suggests she will make a March payment and an October payment which will be 6% of the unpaid balance at that time.

   a. Write a function to describe the interest payments Tammy will pay. Let B represent the unpaid balance from problem 1 and A represent the amount of interest to be paid.

     $A(B) = $ _____

   b. What interest will Tammy pay after the tenth payment? Show work.

3. In any given paragraph, do you think the number of words is a function of the number of sentences? _____ Explain your reasoning.

4. a. Carol's Chocolate Cherries sell for $4.75 per pound. Is the cost of a box of Carol's Chocolate Cherries a function of the weight of the chocolates in pounds? _____
   If so, express the relationship using the function notation and a formula. If not, explain why.

   b. What is the cost of $1\frac{1}{4}$ pounds of Carole's Chocolate Cherries? _____

5. Given the following sequence:

-5	-2	1	4	7	10	...

   a. Predict the next two terms: _____ _____

   b. In the sequence above, $S_3 = $ _____ $S_7 = $ _____ $S_8 = $ _____

   c. Give a rule for the next term in the sequence.

d. Make a table for the first 12 terms in the sequence.

e. Find a pattern or a relationship between each domain number and its image. Once you see the pattern, write a formula for the nth term. (If you have trouble, look at the multiples of 3.)

_____

Why do we use 3 for this pattern?

_____

_____

_____

SOLUTION KEY AND SCORING GUIDE

MATH *Connections* I
How Functions Function Quiz
Sections 6.1 - 6.2 (A)

1. Tammy borrowed $7000.00 from her parents to buy a car. Her parents agreed that she can pay $500.00 on January 1 and $500.00 on July 1 each year to repay the loan.

    a. The first few payments are in the chart below. Complete steps for payments 3, 4 and 5. **(6)**

Payments		Unpaid Balance
		$7000.00
First Payment: Jan. 1	$500.00	$6500.00
Second Payment: July 1	$500.00	$6000.00
Third Payment: Jan. 1	$500.00	$5500.00
Fourth Payment: July 1	$500.00	$5000.00
Fifth Payment: Jan. 1	$500.00	$4500.00

    b. Write a function for the unpaid balance (B) of the loan in terms of the number of loan payments (n).
       B (n) = 7000 - 500n   **(5)**

    c. What is the domain of this function?   {0,1,2,3,. . .14}   **(5)**

    d. What is the range of this function? {7000, 6500, 6000, . . .,0}   **(5)**

    e. What will be the unpaid balance after the 10th payment? _$2000_   **(5)**

    f.   Draw a graph that represents the unpaid balance and the first 5 payments on the loan.
    (9)

    Unpaid Balance

    The graph should be distinct points.

1

2. Tammy offers to pay some interest to her parents on the money she borrowed. She suggests she will make a March payment and an October payment which will be 6% of the unpaid balance at that time.

     a. Write a function to describe the interest payments Tammy will pay. Let B represent the unpaid balance from problem 1 and A represent the amount of interest to be paid. *(5)*

         A(B) = <u> 0.06B </u>

     b. What interest will Tammy pay after the tenth payment? Show work. *(5)*

         A(2000) = 0.06 • 2000 = $120

3. In any given paragraph, do you think the number of words is a function of the number of sentences? <u> No </u> Explain your reasoning. *(6)*

     No. A rich answer would include the statement that the number of sentences does not determine the number of words in a paragraph because not all sentences contain the same number of words.

4.   a. Carol's Chocolate Cherries sell for $4.75 per pound. Is the cost of a box of Carol's Chocolate Cherries a function of the weight of the chocolates in pounds? <u>*Yes*</u> *(3)*
     If so, express the relationship using the function notation and a formula. If not, explain why.

         <u> *C(w) = $4.75 w* </u>   *(5)*

     b. What is the cost of $1\frac{1}{4}$ pounds of Carole's Chocolate Cherries? *(5)*

$$C(\, 1\tfrac{1}{4}\,) = \$4.75(\, 1\tfrac{1}{4}\,) = \$4.75(1.25) = \$5.94$$

5. Given the following sequence:

-5	-2	1	4	7	10	...

     a. Predict the next two terms: <u> *13* </u>   <u> *16* </u>   *(5)*

     b. In the sequence above, $S_3 = $ <u> *1* </u> $S_7 = $ <u> *13* </u> $S_8 = $ <u> *16* </u>   *(6)*

     c. Give a rule for the next term in the sequence. <u> *Add 3* </u>   *(5)*

d. Make a table for the first 12 terms in the sequence. *(9)*

1	-5	7	13
2	-2	8	16
3	1	9	19
4	4	10	22
5	7	11	25
6	10	12	28

e. Find a pattern or a relationship between each domain number and its image.

Once you see the pattern, write a formula for the nth term:

 $f(n) = 3n - 8$ *(5)* (If you have trouble, look at the multiples of 3.)

Why do we use 3 for this pattern? *(6)*

> *The difference between each term is 3 which makes us realize that each domain element was multiplied by 3.*

MATH *Connections* I
How Functions Function Quiz
Sections 6.3 - 6.4 (A)

NAME _____ DATE _____

1. In the past, a bottle of baby aspirin had the following information on the label:

DOSAGE (per day every four hours)	
Under two years of age, use only as directed by a physician	
2 but less than 4 years	2 tablets
4 but less than 6 years	3 tablets
6 but less than 9 years	4 tablets
9 but less than 11 years	5 tablets
11 but less than 12 years	6 tablets
12 or more years	8 tablets

a. Explain why the table above represents a function.

_____

_____

_____

What is the domain?

_____

b. Draw a graph of the function.

c. Discuss why a step-function is appropriate for this situation.

_____

_____

_____

d. For a child $2\frac{1}{2}$ years old, what would be the appropriate dosage?

_____

e. For a child 7 years old, what would be the appropriate dosage?

_____

f. For a child 1 year old, what would be the appropriate dosage? _____

2. a. You may use the graphing calculator Celsius to Fahrenheit program or your Fahrenheit to Celsius program to complete the following table:

City	°F	°C
Hartford CT	38°	
Fairbanks AK		10°
Los Angeles CA	90°	
Bangor ME		0°

b. Your cousin tells you that a certain liquid boils at 100° Celsius. He asks you to find the equivalent Fahrenheit temperature. What is it? _____

3.   a. On the map provided, one inch equals _____ miles.

b. Jean and her friends live in Norwalk and have tickets to the UConn basketball game in Storrs. Jean uses the map to estimate the number of miles from Norwalk to Storrs. What is a good estimate?

_____

c. Write a function for finding the distance between two locations.

_____

d. Jean estimates her average speed will be 50 miles per hour for the trip. How much time should she allow for the trip from Norwalk to Storrs?

_____

4. The freshman class is planning a fund raiser. You are the class treasurer and are responsible for the financial arrangements for the DJ. You have signed a contract with a local disk jockey to play at a Saturday night dance. The DJ charges $150 plus 10% of the total money collected from ticket sales for the dance. You plan to charge $5.00 per ticket. A ticket admits one person.

a. Let *s* represent the number of tickets sold. Let *D* represent the cost for hiring the DJ. Write an equation for *D* as a function for *s*.

_____

b. Graph the function on your graphing calculator. Is it linear? _____ Why?

_____

_____

_____

c. How does the graph help you in making decisions about hiring the DJ? Explain.

_____

_____

_____

d. Find *D* (100) _____ Find *D* (750) _____

e. What does the *y*-intercept represent in this problem?

_____

_____

f. What is the slope? _____ and what does the slope represent in this problem?

_____

_____

g. Let *R* (Revenue) represent the amount of money collected from the sale of tickets before any expenses are deducted (gross sales).

Write a function for *R*. _____

h. Is it linear? _____ What is the domain for this function?

_____

What is the *y*-intercept? _____ What is the slope? _____

i. What is the *x*-intercept? _____ What does the *x*-intercept represent in this problem?

_____

SOLUTION KEY AND SCORING GUIDE

MATH *Connections* I
How Functions Function Quiz
Sections 6.3 - 6.4 (A)

1. In the past, a bottle of baby aspirin had the following information on the label:

DOSAGE (per day every four hours)	
Under two years of age, use only as directed by a physician	
2 but less than 4 years	2 tablets
4 but less than 6 years	3 tablets
6 but less than 9 years	4 tablets
9 but less than 11 years	5 tablets
11 but less than 12 years	6 tablets
12 or more years	8 tablets

a. Explain why the table above represents a function. **(4)**

For each domain value (child's or person's age), there is one range value.

What is the domain? **(4)**

The domain is the set of ages in years..

b. Draw a graph of the function. **(5)**

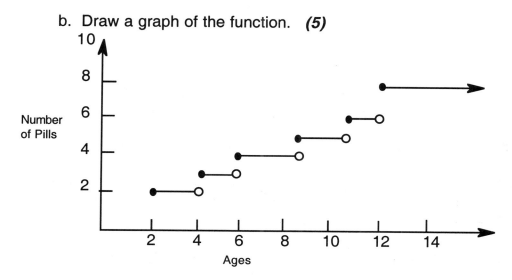

c. Discuss why a step-function is appropriate for this situation. *(4)*

A rich answer should include a discussion that several domain values give only one range value. For every domain value there is only one range value which is a positive integer.

d. For a child $2\frac{1}{2}$ years old, what would be the appropriate dosage? <u>2 tablets</u> *(3)*

e. For a child 7 years old, what would be the appropriate dosage? <u>4 tablets</u> *(3)*

f. For a child 1 year old, what would be the appropriate dosage? <u>undefined</u> *(3)* *(A rich answer should state that for a 1 year old, it would be necessary to get a doctor's opinion for dosage.)*

2. a. You may use the graphing calculator Celsius to Fahrenheit program or your Fahrenheit to Celsius program to complete the following table: *(2 each)*

City	°F	°C
Hartford CT	38°	*3.3°*
Fairbanks AK	*50°*	10°
Los Angeles CA	90°	*32.2°*
Bangor ME	*32°*	0°

b. Your cousin tells you that a certain liquid boils at 100° Celsius. He asks you to find the equivalent Fahrenheit temperature. What is it? <u>*212°*</u> *(4)*

3. a. On the map provided, one inch equals <u>15.4</u> miles. *(4)*

b. Jean and her friends live in Norwalk and have tickets to the UConn basketball game in Storrs. Jean uses the map to estimate the number of miles from Norwalk to Storrs. What would be a good estimate? *(4)*

<u>*5.25inches * 15.4 miles per inch = approximately 80 miles . Students'*</u>
<u>*answers may vary from this as they may measure along highways.*</u>

c. Write a function for finding the distance between two locations. *(4)*
   f(i) = 15.4 i

d. Jean estimates her average speed will be 50 miles per hour for the trip. How much time should she plan for the trip from Norwalk to Storrs? *(4)*

   1 hour and 36 minutes

4. The freshman class is planning a fund raiser. You are the class treasurer and are responsible for the financial arrangements for the DJ. You have signed a contract with a local disk jockey to play at a Saturday night dance. The DJ charges $150 plus 10% of the total money collected from ticket sales. You plan to charge $5.00 per ticket. A ticket admits one person.

   a. Let *s* represent the number of tickets sold. Let *D* represent the DJ's charge. Write an equation for *D* as a function for *s*.
      D(s) = 150 + 10%(5s) = 150 + .10(5 s) (5)

   b. Graph the function on your graphing calculator. Is it linear? *yes* *(2)* Why?

      The points lie on a straight line.

   c. How does the graph help you in making decisions about hiring the DJ? *(4)*

      The graph allows you to quickly determine what the total cost of the DJ will be in terms of the number of tickets that will be sold.

   d. Find *D* (100) *$200.00*   Find *D* (750) *$525.00* *(3 each)*

   e. What does the *y*-intercept represent in this problem? *(3)*

      The y-intercept represents the charge by the DJ if no one attends the dance.

f. What is the slope? __.5__   *(2)*   __and__   what does the slope represent <u>in this</u>
   ___problem?___   *(4)*   *The slope represents the product of the price per ticket*
   and the percent of the ticket sale that goes to the DJ. The DJ receives
   $.50 for each $5.00 ticket that is sold.

g. Let *R* (Revenue) represent the amount of money collected from the sale of
   tickets.
   Write a function for *R.*   __*R = 5s*__
                                                              (4)
h. Is it linear? __*yes*__ *(2)*   What is the domain for this function?   *(4)* __*the*__
   __*number of people who buy tickets; the positive integers*__
   What is the *y*-intercept?_0_   What is the slope? _5_ *(2 each)*

i. What is the *x*-intercept? _0_   *(2)*   What does the *x*-intercept represent
   in this problem?   *(4)* *The x intercept represents the revenue when no one*
   has purchased a ticket.

MATH *Connections* I
How Functions Function Quiz
Section 6.5 (A)

Name_____ Date _____

Recently, Home Savings Bank advertised special saving account rates. You have been given $1000.00 as a gift for your birthday which you want to save. You decide to investigate the different types of accounts available at Home Savings Bank. The rates are listed in the box below:

4.0% interest compounded daily - ordinary passbook account

* 4.25% interest if left in the bank 6 months - 6 month saving account

• 4.5% interest if left in the bank 12 months - 1 year saving account

---

* The interest is compounded at each 6 month interval.
• The interest is compounded at each 12 month interval.

1. Write an equation in the form of $A(n) = P(1 + r)^n$ for each of the three savings plans below. (Note: A(n) represents the amount of money in the bank after the interest has been added, r represents the rate of interest, n represents the number of compounding periods, P represents the amount of money deposited).

   4.0% compounded daily _____

   4.25% compounded every 6 month _____

   4.5% compounded every 12 months _____

2. If you decide to keep the $1000 in the bank for 2 years, determine the amount of money in your account for each plan.

   4.0% compounded daily _____

   4.25% compounded every 6 months _____

   4.5% compounded every 12 months _____

Explain which plan you would choose and why you selected it.

_____

_____

_____

_____

_____

_____

3. If you decide to leave the $1000 in the bank 10 years, determine the amount of money in your account for each plan:

    4.0% compounded daily _____

    4.25% compounded every 6 months _____

    4.5% compounded every 12 months _____

Explain which plan you would choose and why you selected it.

_____

_____

_____

_____

_____

_____

4. If you would like to be able to withdraw money any time, which savings plan would you use? Explain your answer.

_____

_____

_____

_____

_____

_____

SOLUTION KEY AND SCORING GUIDE

MATH *Connections* I
How Functions Function Quiz
Section 6.5 (A)

Recently, Home Savings Bank advertised special saving account rates. You have been given $1000.00 as a gift for your birthday which you want to save. You decide to investigate the different types of accounts available at Home Savings Bank. The rates are listed in the box below:

4.0% interest compounded daily - ordinary passbook account
* 4.25% interest if left in the bank 6 months - 6 month saving account
• 4.5% interest if left in the bank 12 months - 1 year saving account

* The interest is compounded at each 6 month interval.
• The interest is compounded at each 12 month interval.

1. Write an equation in the form of $A(n) = P(1 + r)^n$ for each of the three savings plans below. (Note: A(n) represents the amount of money in the bank after the interest has been added, r represents the rate of interest, n represents the number of compounding periods, P represents the amount of money deposited).*(7 each)*

   4.0% compounded daily $\underline{A(n) = 1000(1 + .04/360)^{360}}$

   4.25% compounded every 6 month $\underline{A(n) = 1000(1 + .0425/2)^2}$

   4.5% compounded every 12 months $\underline{A(n) = 1000(1 + .045)^1}$

2. If you decide to keep the $1000 in the bank for 2 years, determine the amount of money in your account for each plan. *(7 each)*

   4.0% compounded daily $\underline{\$1083.28}$

   4.25% compounded every 6 months $\underline{\$1087.75}$

   4.5% compounded every 12 months $\underline{\$1092.03}$

Explain which plan you would choose and why you selected it. *(13)*

A reasonable answer would be as follows: The plan that yields the most money after two years would be the most logical choice. Therefore the 4.5% compounded yearly would be the choice if you decide to keep the money in the bank for 2 years.

3. If you decide to leave the $1000 in the bank 10 years, determine the amount of money in your account for each plan: *(7 each)*

    4.0% compounded daily *$1491.79*

    4.25% compounded every 6 months *$1522.79*

    4.5% compounded every 12 months *$1552.97*

    Explain which plan you would choose and why you selected it.*(12)*

    The following would be a reasonable answer: The plan that yields the maximum amount over the 10 years would be the logical choice. $4.5% compounded yearly would be the choice in this case.

4. If you would like to be able to withdraw money any time, which savings plan would you use?   Explain your answer. *(12)*

A reasonable answer would include the following points:
a. If a person withdraws money at any point after it is deposited, interest would be applied if the 4.00% plan is chosen.
b. Interest is not applied until 6 months after deposit if the 4.25%, 6 month compounded, plan is chosen, and interest is not applied until 1 year after deposit if the 4.5%, compounded annually, plan is chosen.
c. It makes sense to choose the 4%, compounded daily, plan if you know you plan to withdraw from the account on a regular basis.

MATH *Connections* I
How Functions Function Quiz
Section 6.6 (A)

Name _____   Date _____

Idella Johnson plans to take a trip this summer. She has begun investigating various destinations. The scale of miles on her map is: 1 cm = 50 miles.

1.   a. Write a function, m, to compute the miles represented on the map by a given number of centimeters, c.   _____

      b. Use your function to compute the distances which Idella measured on the map. Record your answers in the table below:

	Map Distance (Centimeters)	Actual Distance (Miles)
Hartford CT to New York City	2.0 cm	_____
Hartford CT to Lexington KY	14. 5 cm	_____
Hartford CT to Chicago IL	15.5 cm	_____
Hartford CT to Denver CO	34.5 cm	_____
Hartford CT to Miami FL	23.0 cm	_____
Hartford CT to San Francisco CA	57.0 cm	_____

2. a. Idella has a new Toyota Camry which averages 35 miles per gallon (mpg) on trips. Write a function, g, which computes the number of gallons to travel m miles. _____

   b. Use your function to compute how many gallons is needed to travel 800 miles. _____

3. a. Write the function g ∘ m: _____

   b. Explain to Idella what the function g ∘ m means to Idella in terms of miles travelled and number of gallons used.

   _____

   _____

   _____

   _____

1

   c. Use g ∘ m to help Idella estimate the amount of gasoline needed for 3 of
      the above trips by car.

   <u>Trip</u>   <u>Gasoline</u>

   _____   _____

   _____   _____

   _____   _____

4. a. Use the average price of regular gasoline as $1.24 and write a function, c, to

       compute the cost of g gallons of gas. _____

    b. Use your function to determine the cost of gas for a trip from Hartford to Miami.
       _____

5. In each of these parts there are formulas for two functions, f and g, on the set of
    rational numbers. In each case,
       • find formulas for g ∘ f and f ∘ g, <u>and</u>

       • use your formulas to compute (g ∘ f)(5) and (f ∘ g)(5)

    a. Given: $f(x) = 3x + 5$ $g(x) = 2x - 1$

       g ∘ f = _____   f ∘ g = _____

       (g ∘ f)(5) = _____   (f ∘ g)(5) = _____

    b. Given: $f(x) = x^2$ $g(x) = 2x - 1$

        g ∘ f = _____   f ∘ g = _____

        (g ∘ f)(5) = _____   (f ∘ g)(5) = _____

SOLUTION KEY AND SCORING GUIDE

MATH *Connections* I
How Functions Function Quiz
Section 6.6 (A)

Idella Johnson plans to take a trip this summer. She has begun investigating various destinations. The scale of miles on her map is: 1 cm = 50 miles.

1.   a. Write a function, m, to compute the miles represented on the map by a given number of centimeters, c.   *m(c) = 50c* **(8)**

   b. Use your function to compute the distances which Idella measured on the map. Record your answers in then table below: **(12)**

	Map Distance (Centimeters)	Actual Distance (Miles)
Hartford CT to New York City	2.0 cm	*100 miles*
Hartford CT to Lexington KY	14. 5 cm	*725 miles*
Hartford CT to Chicago IL	15.5 cm	*775 miles*
Hartford CT to Denver CO	34.5 cm	*1725 miles*
Hartford CT to Miami FL	23.0 cm	*1150 miles*
Hartford CT to San Francisco CA	57.0 cm	*2850 miles*

2.   a. Idella has a new Toyota Camry which averages 35 miles per gallon (mpg) on trips. Write a function, g, which computes the number of gallons to travel m miles. *g(m) = m/35* **(8)**

   b. Use your function to compute how many gallons is needed to travel 800 miles. *g(800) = 800/35 = 22.86 gallons* **(8)**

3.   a. Write the function g ∘ m. *g∘m = 50c / 35* **(4)**

   b. Explain what the function g ∘ m means to Idella in terms of miles travelled and number of gallons used. **(12)**

   g∘m means that Idella can use a map to convert centimeters to number of miles for a trip, then compute the number of gallons of gas that would be used on the trip.

c. Use g ∘ m to help Idella estimate the amount of gasoline needed for 3 of the above trips by car. *(12)*

Trip	Gasoline
Hartford to NYC	*2.86 gallons*
Hartford to Lexington	*20.71 gallons*
Hartford to Chicago	*22.14 gallons*
Hartford to Denver	*49.29 gallons*
Hartford to Miami	*32.86 gallons*
Hartford to SF	*81.43 Gallons*

4.   a. Use the average price of regular gasoline as \$1.24 and write a function, c, to compute the cost of g gallons of gas. ___*c(g) = 1.24g*_____ *(6)*

b. Use your function to determine the cost gas for a trip from Hartford to Miami.
___*c(32.86) = 1.24(32.86) = \$40.75*___ *(6)*

5. In each of these parts there are formulas for two functions, f and g, on the set of rational numbers. In each case,
   - find formulas for g ∘ f and f ∘ g, <u>and</u>

   - use your formulas to compute (g ∘ f)(5) and (f ∘ g)(5) *(3 each)*

a. Given: $f(x) = 3x + 5$ $g(x) = 2x - 1$

g ∘ f = ___*2(3x + 5) - 1 = 6x + 9*___   f ∘ g = ___*3(2x-1) + 5 = 6x + 2*___

(g ∘ f)(5) = __*39*__   (f ∘ g)(5) = __*32*__

b. Given: $f(x) = x^2$ $g(x) = 2x - 1$

g ∘ f = __*2x^2- 1*__   f ∘ g = ___*(2x - 1)^2*___

(g ∘ f)(5) = __*49*__   (f ∘ g)(5) = __*81*___

MATH *Connections* I
How Functions Function Test
Chapter 6 (A)

Name _____ Date _____

1. A bottle of baby Tylenol has the following information on the label:

> ### DOSAGE EVERY 4 HOURS
> Under two years of age, use only as directed by a physician
>
> | | |
> |---|---|
> | 2 but less than 4 years | 1 tablet |
> | 4 but less than 6 years | 2 tablets |
> | 6 but less than 8 years | 3 tablets |
> | 8 but less than 10 years | 4 tablets |
> | 10 but less than 12 years | 6 tablets |
> | 12+ years | 8 tablets |

a. Does the table represent a function? _____ What is the domain?

b. Discuss why a step-function is appropriate for this situation.

_____

_____

_____

_____

c. If a child is $9\frac{1}{2}$ years old, what is the appropriate dosage? _____

d. If a child is 14 years old, what is the appropriate dosage? _____

e. If a child is 4 years old, what is the appropriate dosage? _____

f. Draw a graph of the function.

2. Carol's Chocolate Shoppe sells Chocolate Meltaways for $4.25 per pound, chocolate covered raisins for $3.75 per pound, chocolate covered mixed nuts for $4.00 per pound and white chocolates for $4.00 per pound.

    a. Is the price of the chocolates a function of the type of chocolate? _____ Use a diagram to illustrate why or why not. Label the domain and range.

    b. Is the type of chocolate a function of the price of the chocolates? _____ Use a diagram to illustrate your answer and label the domain and range.

3.

1	2	3	4	5	6	...
January	February	March	April	May	June	

    For the function above:

    a. What is the image of 3? _____

    b. What is the image of 9? _____

    c. Describe this function with a rule.

    d. Is this function a sequence? (Be careful!)

    e. What is the domain of this function? _____
       Is the domain finite? _____

4. One particular kind of billing plan that Southern New England Telephone Company offers charges $0.35 per minute or fraction of minute on local calls.

a. Is this a step function? _____

b. Draw a graph of the function for calls up to 10 minutes in duration.

c. What is the domain?

d. Does it make sense to talk about - 2 minutes? Why or why not?

_____

_____

_____

_____

e. What is the cost of a call lasting 3 minutes 10 seconds?
_____

f. What is the cost of the call lasting 4 minutes? _____

g. What is the cost of the call lasting 2 minutes 55 seconds?
_____

5.

CURRENCY CONVERSION

	Last Week	Prior Week	Year Ago
Japanese Yen per U.S Dollar	105.51	104.80	117.60
German Mark per U.S. Dollar	1.7186	1.7101	1.6698
Canadian Dollar Per U. S. Dollar	1.3571	1.3480	1.2466
British Pound per U. S. Dollar	1.4901	1.4887	1.4460
Gold Republic National Bank	$377.00	$378.50	$329.60

Ahmed is planning a trip to Canada with his French class. The French teacher has been corresponding with a teacher in Canada. In his most recent letter, The Canadian teacher estimated that each student will need 400.00 Canadian dollars to cover living expenses. Ahmed's teacher brought the chart above to class for the students to use to determine how many U.S. dollars they will need.

a. Write a formula for a function that converts Canadian Dollars (c) to U.S. Dollars (A). _____

b. Use your formula to compute how many U.S. Dollars Ahmed needs to save for his Canadian trip.

c. Ahmed's family has a catalog from Canada with handmade woolen items. His mother has asked Ahmed to purchase a heavy woolen shirt from the catalog for his father and she has admired a particular sweater. The price of the shirt is 32.75 Canadian Dollars and the price of the sweater is 48.50 Canadian Dollars. How much must he save in U.S. Dollars in order to buy the shirt for his father? _____ the sweater for his mother? _____

d. Ahmed saved 200.00 in U.S. Dollars for the trip. How much is that in Canadian Dollars? _____ How much more must he save in U.S. Dollars in order to meet his expenses and buy the shirt and sweater? _____

e. If you were Ahmed, how much in U.S. Dollars would you estimate you would need for the trip? _____

7. The total amount of garbage (waste) that is generated by each person in the United States has increased by a factor of 1.1 between both 1970 and 1980 and between 1980 and 1990. The Information Please Environmental Almanac states that each person in the U.S. generates 4.0 pounds of garbage a day.

a. Write a formula for a function G that gives the number of pounds of garbage generated by a person each day for each decade after 1990, assuming the growth rate remains the same.

_____

b. Use the function, G, to compute the approximate number of pounds of garbage generated by each person in the U.S. by 2000 _____ by 2010 _____

c. Put the formula for G in the $\boxed{Y=}$ list of your graphing calculator. Then graph it, setting the $\boxed{window}$ values like this: x from 0 to 10 with a scale of 1 ; y from 0 to 10 with a scale of 1.

d. At this rate of growth, use your graph to approximate how many decades it will take before the number of pounds of garbage each person generates is doubled? _____

Approximately when will that occur? _____

8.   a. Given: $F(x) = 4x$

                   $G(x) = 2x - 1$

       Determine: F o G   or   F(G(x)) _____

       Determine: G o F   or   G(F(x)) _____

       Determine the values of: F(2) _____ G(2) _____

                             F(G(2)) _____   G(F(2)) _____

       b. Given: $t(x) = x\,2$

                   $h(x) = 3x - 1$

       Determine: t o h or   t(h(x)) _____

       Determine: h o t   or   h(t(x)) _____

       Determine the values of: t(-3) _____   h(-3) _____

                             t(h(-3)) _____ h(t(-3)) _____

SOLUTION KEY AND SCORING GUIDE

MATH *Connections* **I**
How Functions Function Test
Chapter 6 (A)

1. A bottle of Tylenol has the following information on the label:

DOSAGE EVERY 4 HOURS	
Under two years of age, use only as directed by a physician	
2 but less than 4 years	1 tablets
4 but less than 6 years	2 tablets
6 but less than 8 years	3 tablets
8 but less than 10 years	4 tablets
10 but less than 12 years	6 tablets
12+ years	8 tablets

a. Does the table represent a function? _yes_ **(1)** What is the domain?
                                                        (2)

The domain is the ages of people.

b. Discuss why a step-function is appropriate for this situation. **(3)**

The answer should include the idea that several intervals of the domain determine the same range value over the interval. The graph in these domains is a horizontal line which produces a step-like graph. For example for ages 4 -5, the number of tablets is 3.

c. If a child is $9\frac{1}{2}$ years old, what is the appropriate dosage? _4 tablets_ **(2)**

d. If a child is 14 years old, what is the appropriate dosage? _8 tablets_ **(2)**

e. If a child is 4 years old, what is the appropriate dosage? _2 tablets_ **(2)**

f. Draw a graph of the function. *(5)*

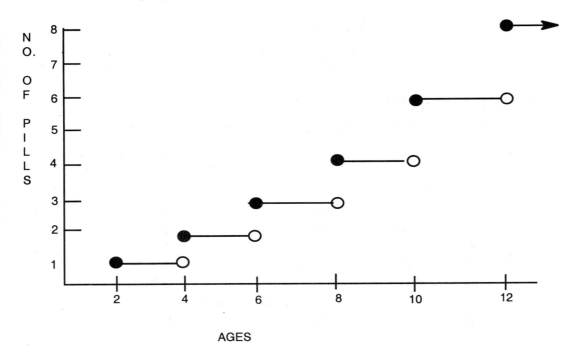

AGES

2. Carol's Chocolate Shoppe sells Chocolate Meltaways for $4.25 per pound, chocolate covered raisins for $3.75 per pound, chocolate covered mixed nuts for $4.00 per pound and white chocolates for $4.00 per pound.

    a. Is the price of the chocolates a function of the type of chocolate? __yes__ *(1)* Use a diagram to illustrate your answer and label the domain and range. *(2)*

Candy	Price
C. M.	$4.25
C.C.R.	$3.75
C.C.M.N.	$4.00
W.C.	$4.00

    The Candy is the domain; the price is the range. It is a function as each domain value has a unique range value.

    b. Is the type of chocolate a function of the price of the chocolates? __no__ *(1)* Again, use a diagram to illustrate why or why not. Label the domain and range. *(2)*

Price	Candy
4.25	C.M
3.75	C.C.R.
4.00	C.C.M.N.
4.00	W.C.

2

The domain is the price; the range is the candy. It is not a function as the same domain value produces different range values.

3.

1	2	3	4	5	6	...
January	February	March	April	May	June	

For the function above: *(2 each)*

a. What is the image of 3? *March*

b. What is the image of 9? *September*

c. Describe this function with a rule. *The whole numbers from 1 to 12 are matched with the months of the year in order.*

d. Is this function a sequence? Be careful! *No. The range values must be numbers.*

e. What is the domain of this function? *The whole numbers from 1 to 12*
    Is the domain finite? *yes*   *(1)*

4. One particular kind of billing plan that Southern New England Telephone Company offers charges $.035 per minute or fraction of minute on local calls.

a. Is this a step function ? *yes*   *(1)*

b. Draw a graph of the function up to 10 minutes. *(4)*

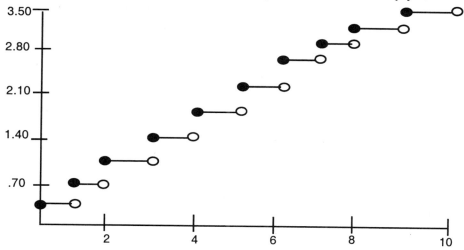

c. What is the domain? *The positive rational numbers.* *(2)*

d. Does it make sense to talk about - 2 minutes? Why or why not? *No, because time is not measured in negative units.* *(3)*

e. What is the cost if a call lasts 3 minutes 10 seconds? *$1.40*   *(2)*

f. What is the cost if the call lasts 4 minutes? *$1.40*   *(2)*

g. What is the cost if the call lasts 2 minutes 55 seconds?   *$1.05*   *(2*

5.

CURRENCY CONVERSION			
	Last Week	Prior Week	Year Ago
Japanese Yen per U.S Dollar	105.51	104.80	117.60
German Mark per U.S. Dollar	1.7186	1.7101	1.6698
Canadian Dollar Per U. S. Dollar	1.3571	1.3480	1.2466
British Pound per U. S. Dollar	1.4901	1.4887	1.4460
Gold Republic National Bank	$377.00	$378.50	$329.60

Ahmed is planning a trip to Canada with his French class. The French teacher has been corresponding with a teacher in Canada. In his most recent letter, The Canadian teacher estimated that each student will need $400.00 Canadian dollars to cover living expenses. Ahmed's teacher brought the chart above to class for the students to use to determine how many U.S. dollars they will need..

   a. Write a formula for a function that converts Canadian Dollars (c) to
       U.S. Dollars (A). *US$ = 1.3571 * C$*   *(3)*

   b. Use your formula to compute how many U.S. Dollars Ahmed needs
       to save for his Canadian trip. *(2)*

       *US$ = 1.3571 * 400 = $542.84*

c. Ahmed's family has a catalog from Canada with handmade woolen items. His mother has asked Ahmed to purchase a heavy woolen shirt from the catalog for his father and she has admired a particular sweater. The price of the shirt is 32.75 Canadian Dollars and the price of the sweater is 48.50 Canadian Dollars. How much must he save in U.S. Dollars in order to buy the shirt for his father? _$44.45_ *(2)* the sweater for his mother? _$65.82_ *(2)*

d. Ahmed saved $200.00 in U.S. Dollars for the trip. How much is that in Canadian Dollars? _147.37_ *(2)* How much more must he save in U.S. Dollars in order to meet his expenses and buy the shirt and sweater? _$505.74_ *(2)*

e. If you were Ahmed, how much in U.S. Dollars would you estimate you would need for the trip? _$700.00_ *An acceptable answer would be be that Ahmed could use between $700 and $800 to allow for some spending money.* *(2)*

7. The total amount of garbage (waste) that is generated by each person in the United States has increased by a factor of 1.1 between both 1970 and 1980 and between 1980 and 1990. The Information Please Environmental Almanac" states that each person in the U.S. generates 4.0 pounds of garbage a day.

a. Write a formula for a function G that gives the amount of garbage generated by a person each day for each decade after 1990, assuming the growth rate remains the same.

_G(X) = 4(1.1)_^{X} *(3)*

b. Use the function, G, to compute the approximate garbage generated by each person in the U.S. by 2000 _4.4 pounds_ *(2)* by 2010 _4.84 pounds_ *(2)*

c. Put the formula for G in the $\boxed{Y=}$ list of your graphing calculator. Then graph it, setting the $\boxed{\text{window}}$ values like this: x from 0 to 10 with a scale of 1 ; y from 0 to 10 with a scale of 1.

d. At this rate of growth, use your graph to approximate how many decades it will take before the amount of garbage each person generates is doubled?

      7 decades   **(2)**

Approximately when will that occur?   *between 2065 and 2066*   **(2)**

8.   a. Given: $F(x) = 4x$

                   $G(x) = 2x - 1$ **(2 each)**

Determine: F o G   or   F(G(x))   *4(2x - 1) or 8x - 4*

Determine: G o F   or   G(F(x))   *2(4x) - 1 or 8x - 1*

Determine the values of: F(2) *8* G(2) *3*

                               F(G(2)) *12*   G(F(2)) *15*

b. Given: $t(x) = x^2$

                 $h(x) = 3x - 1$ **(2 each)**

Determine: t o h   or   t(h(x)) *(3x - 1)²*

Determine: h o t   or   h(t(x)) *3x² -1*

Determine the values of: t(-3) *9*   h(-3) *- 10*

                          t(h(-3)) *100*   h(t(-3)) *26*

MATH *Connections* I
Counting Beyond 1, 2, 3 Quiz
Sections 7.1 - 7.2 (A)

Name _____ Date _____

1. A survey of the weights of students in the Grade 9 class at Park High School is
   being conducted by a science class . The lightest student weighs 105 pounds and
   the heaviest weighs 192 pounds. In order to simplify computer input, it is decided to
   code the weights in intervals of 5. Design a coding system that follows this rule.
   Explain why you chose your coding system.

   _____

   _____

   _____

   _____

   _____

   _____

   _____

2. A computer at United Technologies does 50,000 operations per second.   A
   problem involving aerodynamics is put into the computer. It requires 10^{10} operations
   to complete the problem. Your job is to watch the computer and report when the
   problem has been completed. The computer starts at 8:00 am.   Will you need to
   send out for lunch? _____ Explain.

   _____

   _____

   _____

   _____

   _____

   _____

3. a. List the different orders in which Roberto, Charles and Huong can sit in three
       chairs in a row.

    b. If a fourth chair is provided and Tim joins the group, how many different orders
        can be made? _____

4. Describe the set of two-digit numbers that contain a 2 by making a list.

5. If set G represents the set of two-digit numbers that contain the number 2, what is
    #G? _____

6. a. If you roll a six-sided cube numbered 1 to 6 and then flip a dime, list the set of all
       possible outcomes (P).

    b. What is #(P)? _____

    c. You roll a ten-sided polyhedron numbered 1 to 10 and then flip a dime. The set
        of possible outcomes is named T. What is #(T)? _____

SOLUTION KEY AND SCORING GUIDE

MATH *Connections* I
Counting Beyond 1, 2, 3 Quiz
Sections 7.1 - 7.2 (A)

1. A survey of the weights of students in the Grade 9 class at Park High School is being conducted by a science class . The lightest student weighs 105 pounds and the heaviest weighs 192 pounds. In order to simplify computer input, it is decided to code the weights in intervals of 5. Design a coding system that follows this rule. Explain why you chose your coding system. *(18)*

The response should include a plan to include all the possible weights in intervals of 5 from 105 to 192. There are many different ways this could be accomplished. One example is : 105 - 109; 110 - 114; 115 - 119; 120 - 124; 125 - 129; 130 -134; 135-139; 140 - 144; 145-149; 150-154; 155-159; 160-164; 165-169; 170-174; 175-179; 180-184; 185-189; 190 -194.

2. A computer at United Technologies does 50,000 operations per second. A problem involving aerodynamics is put into the computer. It requires 10^{10} operations to complete the problem. Your job is to watch the computer and report when the problem has been completed. The computer starts at 8:00 am. Will you need to send out for lunch? __yes__ *(5)* Explain. *(12)*

The explanation should contain the information that it will take 200,000 seconds or 333.33... minutes or 55.5... hours. This would mean that you would have to send out for lunch.

3. a. List the different orders in which Roberto, Charles and Huong can sit in three chairs in a row. *(15)*

   R,C,H   R,H,C   C,R,H   C,H,R   H,R,C   H,C,R

   b. If a fourth chair is provided and Tim joins the group, how many different orders can be made? *(5)* 24

4. Describe the set of two-digit numbers that contain 2 by making a list. *(10)*

   {12, 20, 21, 22, 23, 24, 25, 26, 27, 28, 29, 32, 42, 52, 62, 72, 82, 92}

5. If set G represents the set of two-digit numbers that contain the number 2, what is #G?   *(5)*

    _18_

6. a. If you roll a six-sided cube numbered 1 to 6 and then flip a dime, list the set of possible outcomes (P). *(15)*

    {(1,H), (2,H), (3,H), (4,H), (5,H), (6,H), (1,T), (2,T), (3,T), (4,T), (5,T), (6,T)}

    b.   What is #(P)? *(5)*   _12_

    c. You roll a ten-sided polyhedron numbered 1 to 10 and then flip a dime. The set of possible outcomes is named T. What is #(T)?   *(10)*

    _20_

MATH *Connections* I
Counting Beyond 1, 2, 3 Quiz
Section 7.3 (A)

Name _____ Date _____

Directions for examples 1, 2 and 3:
   a. Shade the given Venn diagram to illustrate the relationship between the two sets.

   b. Write a sentence to explain the relationship.

1.   A ∩ B

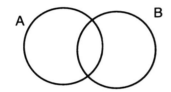

2.   A ∪ B

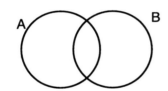

3.   A – B

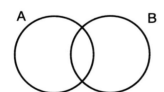

4. Draw a Venn diagram to illustrate the relationship between the two sets <u>and</u> write a sentence to explain each of the following .
   a. A and B are disjoint sets.   b. A $\subset$ B

5. a. Demonstrate by shading: (A $\cup$ B) $\cup$ C and A $\cup$ (B $\cup$ C)

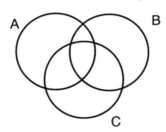   

   b. What concept does your shading demonstrate? _____

6. Students in the Connections class at Central High School filled out a survey on their favorite bands.   12 students selected Aerosmith, 15 selected Bon Jovi and 8 selected both.   6 students chose other bands.
   a. Draw a Venn diagram to illustrate this information.

   b. How many students are in the class?_____

7. Ms Hamilton's homeroom has 18 girls. 10 girls enjoy playing tennis or softball,
    5 enjoy playing tennis only, and 3 enjoy playing softball only.
    a. Draw a Venn diagram to illustrate this information.

    b. How many enjoy playing both tennis and softball? ___

8. Fifty-six people signed the Declaration of Independence. Forty people signed the
   U.S. Constitution. Five people signed both documents.
    a. Draw a Venn diagram to illustrate this information.

    b. How many people signed the Constitution <u>or</u> the Declaration of Independence
    <u>but not</u> both ? _____

9. The Proudankle Athletic Shoe Company manufactures 27 different models of
    athletic shoes.   Ten models of shoes are recommended for running, 8 models are
    recommended for aerobics, 7 models are recommended for walking while 6
    models are recommended for both aerobics and walking.
    a. Draw a Venn diagram to illustrate this information.

    b. How many models are manufactured that are not recommended
       for running, aerobics or walking? _____

    c. How many choices of shoes are there for people who do aerobics
       or walking? _____

SOLUTION KEY AND SCORING GUIDELINES

MATH *Connections* I
Counting Beyond 1, 2, 3 Quiz
Section 7.3 (A)

Directions for examples 1, 2 and 3:
  a. Shade the given Venn diagram to illustrate the relationship between the two sets.

  b. Write a sentence to explain the relationship. **(5 each)**

1. A ∩ B

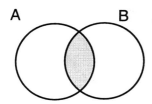

The intersection of two sets are the elements they have in common.

2. A ∪ B

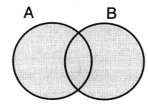

The union of two sets is all the elements contained in one set or the other or both.

3. A − B

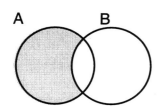

The difference between A and B is the set of elements in A but not in B.

1

4. Draw a Venn diagram to picture the relationship between the two sets <u>and</u>
   write a sentence to explain each of the following . *(5 each)*
   a. A and B are disjoint sets.   b. A ⊂ B

   *A and B are disjoint if they   A is a subset of B if all the elements of
       have no common elements.   A are contained in B.*

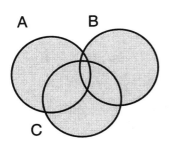

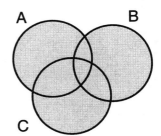

5. a. Demonstrate by shading: (A ∪ B) ∪ C and A ∪ (B ∪ C)   *(5)*

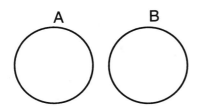

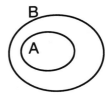

   b. What concept does your shading demonstrate? *(5)*

   *The illustration demonstrates that the operation of taking the union is
   associative.*

6. Students in the Connections class at Central High School filled out a survey
   on their favorite bands. 12 students selected Aerosmith, 15 selected Bon Jovi and
   8 selected both. 6 students chose other bands.
   a. Draw a Venn diagram to illustrate this information. *(10)*

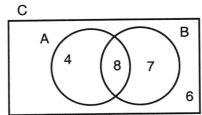

   b. How many students are in the class? _25_   *(5)*

7. Ms Hamilton's homeroom has 18 girls. 10 girls enjoy playing tennis or softball, 5 enjoy playing tennis only, and 3 enjoy playing softball only.
   a. Draw a Venn diagram to illustrate this information. *(10)*

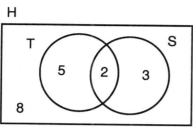

   b. How many enjoy playing both tennis and softball? _2_ *(5)*

8. Fifty-six people signed the Declaration of Independence. Forty people signed the U.S. Constitution. Five people signed both documents.
   a. Draw a Venn diagram to illustrate this information. *(10)*

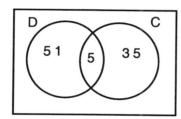

   b. How many people signed the Constitution or the Declaration of Independence but not both ? *(5)*   _86_

9. The Proudankle Athletic Shoe Company manufactures 27 different models of athletic shoes. 10 models of shoes are recommended for running, 8 models are recommended for aerobics, 7 models are recommended for walking while 6 models are recommended for both aerobics and walking.
   a. Draw a Venn diagram to illustrate this information. *(10)*

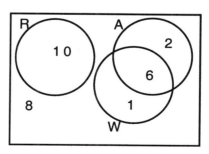

   b. How many models are manufactured that are not recommended for running, aerobics or walking? *(5)*   _8_

   c. How many choices of shoes are there for people who do aerobics or walking? *(5)*   _9_

MATH *Connections* I
Counting Beyond 1, 2, 3 Quiz
Sections 7.4 - 7.5 (A)

Name _____ Date _____

1. Tom, Andy, Marcia and Joan are in a singles badminton tournament and will all play against each other.

   a. Use a tree diagram to show all the possible orders in which Tom could play against Andy, Marcia and Joan.

   b. How many different orders are there for Tom? _____
   c. How many different orders are there for Andy? _____
   d. What is the total number of different orders? _____

2. Pedro's Screaming Ice Cream has a special on ice cream sundaes. You may choose one flavor of ice cream, one sauce and one topping for $1.99. Your choices of ice cream are chocolate chip, fudge swirl and butter crunch; your choices of sauce are fudge, butterscotch and marshmallow; your choices of topping are M&M's, Butterfingers, Heath Bar and Nutty Mix.

   a. Use a tree diagram to illustrate the different sundaes that could be made if you chose **chocolate chip** as the ice cream.

b. How many different sundaes could be made if all the ice creams, sauces and toppings could be selected? _____

c. Which one would you order? _____

3. Marilyn has three projects to complete in her science class. Each project must be from one of the following three areas of science: biology, chemistry or physics. Her teacher has chosen 7 different biology topics, 8 different chemistry topics and 5 different physics topics.

If Marilyn must do one topic from each science area, how many different sets of projects could she do? _____

4. Northern New England University has 4350 students. Each student needs a different identification code for use in the library. Mr. Connors in the data processing center decides that a five character code made up of 2 letters followed by 3 digits is the best way to go. No letter or digit can be repeated and zeroes will not be used.

a. Could he make the code shorter than 5 elements or should he make it longer? _____

Explain.

_____

_____

_____

b. If the code must be made shorter or longer, explain how you would redesign it and why you decided to choose the design.

_____

_____

_____

5. Hamal is going to buy a new car. He chooses the XT-20 model. He must select from the following options: 5 exterior colors; 3 interior colors; manual or automatic transmission; 4-cylinder or 6-cylinder; designer hubcaps or standard hubcaps.

How many different versions of the XT-20 model could he select? _____

Show how you determined your answer.

_____

_____

_____

SOLUTION KEY AND SCORING GUIDE

MATH *Connections* I
Counting Beyond 1, 2, 3 Quiz
Sections 7.4 - 7.5 (A)

1. Tom, Andy, Marcia and Joan are in a singles badminton tournament and will all play against each other.

   a. Use a tree diagram to show all the possible orders in which Tom could play against Andy, Marcia and Joan. *(12)*

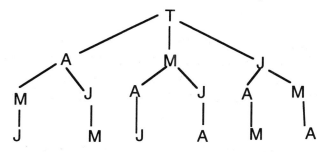

   b. How many different orders are there for Tom? __6__ *(3)*
   c. How many different orders are there for Andy? __6__   *(3)*
   d. What is the total number of orders? __24__   *(3)*

2. Pedro's Screaming Ice Cream has a special on ice cream sundaes. You may choose one flavor of ice cream, one sauce and one topping for $1.99. Your choices of ice cream are chocolate chip, fudge swirl and butter crunch; your choices of sauce are fudge, butterscotch and marshmallow; your choices of topping are M&M's, butterfingers, Heath Bar and nutty mix.

   a. Use a tree diagram to illustrate the different sundaes that could be made if you chose **chocolate chip** as the ice cream.

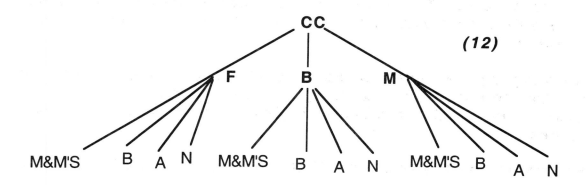

(12)

b. How many different sundaes could be made if **all** the ice creams, sauces and toppings could be selected? _36_   *(8)*

c. Which one would you order? *Answers will vary.*   *(3)*

3. Marilyn has three projects to complete in her science class. Each project must be from one of the following three areas of science: biology, chemistry or physics. Her teacher has chosen 7 different biology topics, 8 different chemistry topics and 5 different physics topics.

   If Marilyn must do one topic from each science area, how many different sets of projects could she do? _7 x 8 x 5 = 280_ *(12)*

4. Northern New England University has 4350 students. Each student needs a different identification code for use in the library. Mr. Connors in the data processing center decides that a five character code made up of 2 letters followed by 3 digits is the best way to go. No letter or digit can be repeated and zeroes will not be used.

   a. Could he make the code shorter than 5 elements or should he make it longer?
          shorter
      Explain. *(12)*

      To determine the number of codes you would multiply 26 x 25 x 9 x 8 x 7 = 327,600. This is far more than the number of codes that are needed.

   b. If the code must be made shorter or longer, explain how you would redesign it and why you decided to choose the design.
      (12)

      He has far more possibilities than he needs. He could use less digits, for example 2 letters and one digit, or he could use 1 letter and 3 digits and have enough codes. The answer should demonstrate some experimenting with possible codes to produce the number of codes necessary for the entire student body.

5. Hamal is going to buy a new car. He chooses the XT-20 model. He must select from the following options: 5 exterior colors; 3 interior colors; manual or automatic transmission; 4-cylinder or 6-cylinder; designer hubcaps or standard hubcaps.

   How many different versions of the XT-20 model could he select? _120_ *(10)*

   Show how you determined your answer. _5 x3 x 2 x 2 x 2 = 120_   *(10)*

MATH *Connections* I
Counting Beyond 1,2,3 Test
Chapter Seven (A)

Name _____   Date _____

1. Given: A = {Al, Andy, Art}
              B = {Al, Art, Bob, Betty}
              C = {Bob, Carol, Cal, Chloe, Cassie}

   a. Write set A in set builder notation.

   b. What does the notation A ∪ B represent?   _____

   Draw a Venn diagram to illustrate the elements in A ∪ B.

   c. What does the notation B ∩ C represent? _____

   Draw a Venn diagram to illustrate the elements of   B ∩ C.

   d. What does the notation A ∩ C represent?_____

   Draw a Venn diagram to illustrate the elements of A ∩ C.

1

e. What is #(B)? _____

2. A survey was held in the East End shopping mall. The Survey included the following statements:

☐   I like winter.

☐   I like summer.

The survey showed: 183 people liked winter.   Only 88 people liked summer.   Only 59 people liked both winter and summer.   102 people did not check either of the two boxes.

a. Display the results by using a Venn Diagram.

b. How many people liked summer? _____

c. How many people liked winter? _____

d. How many people liked either summer or winter or both? _____

e. What conclusions do you draw from the survey?

3. There are 1500 students at Brooklawn High School.   Identification numbers (ID's) are needed for the students. It is decided to give each student an ID which consists of 2 letters followed by a 2-digit number. The letter O and the number 0 cannot be used and the letters and numbers cannot be repeated.

Will there be enough different ID's for the students at Brooklawn High School? Explain your answer.

4. The PowerPlus Automobile offers many choices in colors and styles. The options are listed below:

Exterior Color	Interior Color	Body Style
Red	Maroon	4-door Sedan
Raspberry	White	2-door Coupe
Black	Grey	Convertible
Aqua	Neutral	

Entertainment
Stereo: AM/FM
Stereo: AM/FM with Cassette
Stereo: AM/FM with Compact Disc
Stereo: AM/FM with Cassette and Compact Disc

Seats
Leather
Cloth

a. What is the maximum number of different cars that could be ordered? _____

b. Choose your car and describe your selection:

5. Thirty two basketball teams will play a single elimination tourney (lose one game and your team is eliminated from the tourney) to decide the Class L champion.

a. Make a tree diagram to show how the tourney champion would be decided.

b. How many games are played? _____

c. How many games must the champion win? _____

d. How many games are played if there are 16 teams? ____
   64 teams? _____

e. How would you organize a 16 team tournament so that there is a possibility for the first and second seeded (the two 'best') teams to play in the final game?

6. Mrs. Howard's Connections class decides to change the seats of the five students sitting in the front rows every day until every possible arrangement of the five has been used.

a. How many days will it take until all the arrangements have been used? _____

b. Demonstrate how you determined your answer.

SOLUTION KEY AND SCORING GUIDE
MATH *Connections* **I**
Counting Beyond 1,2,3 Test
Chapter Seven (A)

1. Given: A = {Al, Andy, Art}
           B = {Al, Art, Bob, Betty}
           C = {Bob, Carol, Cal, Chloe, Cassie}

   a. Write set A in set builder notation.
      A = {x | x are names of boys beginning with the letter A}

   b. What does the notation A ∪ B represent?   *union; {Al,Andy,Art,Bob,Betty}*

   Draw a Venn diagram to illustrate the elements in A ∪ B.

   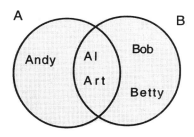

   c. What does the notation B ∩ C represent? *intersection; {Bob}*

   Draw a Venn diagram to illustrate the elements of B ∩ C.

   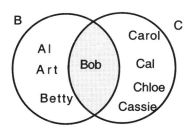

   d. What does the notation A ∩ C represent? *intersection; none*

   Draw a Venn diagram to illustrate the elements of A ∩ C.

   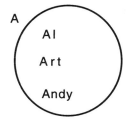 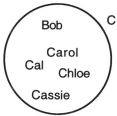

e. What is #(B)? _4_

2. A survey was held in the East End shopping mall. The Survey included the following statements:

☐   I like winter.

☐   I like summer.

The survey showed: 183 people liked winter only.   Only 88 people liked summer only.   Only 59 people liked both winter and summer.   102 people did not check either of the two boxes.

a. Display the results by using a Venn Diagram.

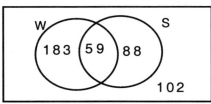

b. How many people liked summer?   _147_

c. How many people liked winter? _242_

d. How many people liked either summer or winter or both? _330_

e. What conclusions do you draw from the survey?

Answers will vary.

3. There are 1500 students at Brooklawn High School. Identification numbers (ID's) are needed for the students. It is decided to give each student an ID which consists of 2 letters followed by a 2-digit number. The letter O and the number 0 cannot be used and the letters and numbers cannot be repeated.

Will there be enough different ID's for the students at Brooklawn High School? Explain your answer.

25 • 24 • 9 • 8 = 43 200

Yes. There are enough ID numbers to last many years without duplication.

2

4. The PowerPlus Automobile offers many choices in colors and styles. The options are listed below:

Exterior Color
Red
Raspberry
Black
Aqua

Interior Color
Maroon
White
Grey
Neutral

Body Style
4-door Sedan
2-door Coupe
Convertible

Entertainment
Stereo: AM/FM
Stereo: AM/FM with Cassette
Stereo: AM/FM with Compact Disc
Stereo: AM\ /FM with Cassette and Compact Disc

Seats
Leather
Cloth

a. What is the maximum number of different cars that could be ordered?
   4 • 4 • 3 • 4 • 2 = 384

b. Choose your car and describe your selection:

   Answers will vary.

5. Thirty two basketball teams will play a single elimination tourney (lose one game and your team is eliminated from the tourney) to decide the Class L champion.

a. Make a tree diagram to show how the tourney champion would be decided.

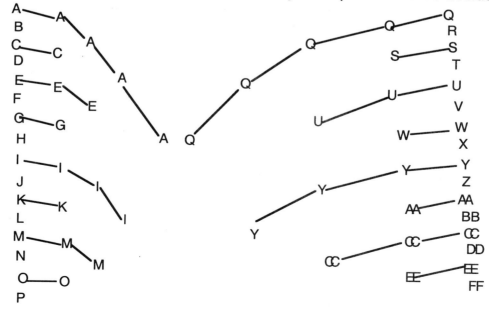

b. How many games are played? __31__

c. How many games must the champion win? __5__

d. How many games are played if there are 16 teams? __15__
   64 teams? __63__

e. How would you organize a 16 team tournament so that there is a possibility for the first and second seeded (the two 'best') teams to play in the final game?

The 16 teams would be divided in half and the first seeded team would be in one half and the second seeded team would be in the other half.

6. Mrs. Howard's Connections class decides to change the seats of the five students sitting in the front rows every day until every possible arrangement of the five has been used.

a. How many days will it take until all the arrangements have been used? __120__

b. Demonstrate how you determined your answer.

$$5 \cdot 4 \cdot 3 \cdot 2 \cdot 1 = 120$$

MATH *Connections* I
What are the Chances? Quiz
Sections 8.1 - 8.2 (A)

Name _____   Date _____

1. • Select a number on the number line which indicates the probability you would
     assign to the following events:
   • Explain your reasoning for each.

   a. I will sleep sometime tonight.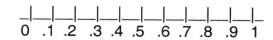

   _____

   _____

   _____

   b. I will throw a snowball today.

   |—|—|—|—|—|—|—|—|—|—|
   0 .1 .2 .3 .4 .5 .6 .7 .8 .9 1

   _____

   _____

   _____

   c. I will see my sister today.

   |—|—|—|—|—|—|—|—|—|—|
   0 .1 .2 .3 .4 .5 .6 .7 .8 .9 1

   _____

   _____

   _____

2. Describe an event (E) that has a probability of $\frac{1}{2}$ i.e. (P(E) = $\frac{1}{2}$).

3. A friend challenges you to a game. He has a cup containing 3 marbles. Two marbles are white and one is blue. He asks you to blindly select one marble and if it is blue, you win, and he will treat you to a video game at the local video parlor. Otherwise, you will treat him to a free video game.

     a. What are the chances for you to win a free video game?

     b. Would you play this game? Explain why or why not.

_____

_____

_____

4. You have a six-sided cube which has the following numbers on its faces: 2, 4, 6, 8, 10, 12. If you roll the cube exactly once, assign probabilities to the following:

     a. P(odd number): _____

     b. P(even number): _____

     c. P(number greater than 10) _____

5. Explain what is meant when we talk about the probability of something happening. Give an example.

_____

_____

_____

_____

_____

6. The eight letters of the word "OKLAHOMA" are written on slips and placed in a bowl. The slips are thoroughly mixed and one of them is picked at random.

     a. What is the probability that an "O" is picked? _____
     b. What is the probability that a "T" is picked? _____
     c. What is the probability that a consonant is picked? _____
     d. What is the probability that the picked letter is in the word "MINNESOTA"? _____

SOLUTION KEY AND SCORING GUIDE

MATH *Connections* **I**
What are the Chances? Quiz
Sections 8.1 - 8.2 (A)

1. • Select a number on the number line which indicates the probability you would assign to the events:
   • Explain your reasoning for each.   *(10 each)*

   Answers will vary. Purpose of questions is to diagnose the student's understanding of the numerical assignment of probabilities.

   a. I will sleep sometime tonight.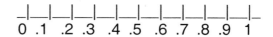

   b. I will throw a snowball today.

   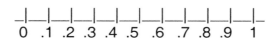

   c. I will see my sister today.

   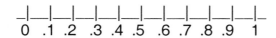

2. Describe an event (E) that has a probability of $\frac{1}{2}$, i.e. (P(E) $=\frac{1}{2}$). *(10)*

   The answers to this question will vary. It is expected that the student would use flipping a coin, the birth of a male or female as prime examples.

3. A friend challenges you to a game. He has a cup containing 3 marbles. Two marbles are white and one is blue. He asks you to blindly select one marble and if it is blue, you win, and he will treat you to a video game at the local video parlor. Otherwise, you will treat him to a free video game.
   a. What are the chances for you to win a free video game? *(10)*

   The chances are not good as there is only a 1 out of 3 probability to select the blue marble on the first draw.

b. Would you play this game? Explain why or why not. *(5)*

The game is not a fair one because of the small probability to win is $\frac{1}{3}$ and the probability to lose is $\frac{2}{3}$. It would be expected that the student would select not to play the game.

4. You have a six-sided cube which has the following numbers on its faces: 2, 4, 6, 8, 10, 12. If you roll the cube exactly once, assign probabilities to the following: *(7 each)*

   a. P(odd number): __0__

   b. P(even number): __1__

   c. P(number greater than 10) __$\frac{1}{6}$__

5. Explain what is meant when we talk about the probability of something happening. Give an example. *(16)*

   A rich answer would include a discussion that the probability of something happening means the ratio of the success of the outcome compared to all possible outcomes, successful and not successful. The student could mention that probabilities range from 0 to 1, inclusive. The student could also use the ratio: successful outcomes / total outcomes in the sample spaces.
   The examples of a probability situation will vary.

6. The eight letters of the word "OKLAHOMA" are written on slips and placed in a bowl. The slips are thoroughly mixed and one of them is picked at random. *(8)*

   a. What is the probability that an "O" is picked? __$\frac{1}{4}$__
   b. What is the probability that a "T" is picked? __0__
   c. What is the probability that a consonant is picked? __$\frac{1}{2}$__
   d. What is the probability that the picked letter is in the word "MINNESOTA"?__$\frac{5}{8}$__

MATH *Connections* **I**
What are the Chances? Quiz
Section 8.3 (A)

Name _____ Date _____

1. You have a six-sided cube numbered 2, 4, 6, 8, 10, 12.
    Write the sample space for rolling the cube once.

2. You roll two cubes which are numbered the same as in problem 1.
    a. A table describing the sample space for **the sums** when   rolling
        these two cubes is started below. Please complete the table.

+	2	4	6	8	10	12
2					12	
4				12		
6						
8	10					
10						
12		16				

b. What is the probability of obtaining a sum of 14? _____

c. What is the probability of obtaining a sum of 24? _____

d. What is the probability of obtaining a sum of 11? _____

e. What is the probability of obtaining an even sum? _____

f. What is the probability of obtaining a sum of 16 and at least one
    cube shows a 6? _____

3. If you were designing a fair game using eight-sided solids, what would be your directions?

    _____

    _____

    _____

    _____

    _____

4. There are 23 students in Mr. Jones' physical education class. 10 of the students are boys and 13 are girls.
    If Mr Jones randomly chooses one person to set up the activity for the day, what is the probability that:

    a. the person will be a girl? _____

    b. the person will be a boy?   _____

    c. the person will not be a girl?   _____

    d. Using the idea of complement, explain the relationship of your answer to a to your answer to c.

    _____

    _____

    _____

5. The electricity is out and Mary has to get up in the dark and dress for work. She has a green, a blue and a white blouse; a black and a red skirt; a pink, a white and a tan sweater. If Mary chooses a blouse, skirt and sweater without seeing their color, what is the probability that she will be wearing a white blouse, a black skirt and a pink sweater?

2

6. Students in the Connections class at Central High School filled out a survey on their favorite bands.   12 students selected Aerosmith, 15 selected Bon Jovi and 8 selected both.   6 students chose other bands.
   a. Draw a Venn diagram to illustrate this information.

   b. How many students are in the class?_____

   c. What is the probability that a student selected at random selected Bon Jovi or Aerosmith, but not both?   _____

   d. What is the probability that a student did not select at random Bon Jovi only?

   _____

   e. Explain how the concept of complement could have been used to solve question d.

   _____

   _____

   _____

   _____

SOLUTION KEY AND SCORING GUIDE

MATH *Connections* I
What are the Chances? Quiz
Section 8.3 (A)

1. You have a six-sided cube numbered 2, 4, 6, 8, 10, 12.
   Write the sample space for rolling the cube once. **(6)**

   {2, 4, 6, 8, 10, 12 }

2. You roll two cubes which are numbered the same as in problem 1.
   a. A table describing the sample space for **the sums** when rolling
   these two cubes is started below. Please complete the table.**(10)**

+	2	4	6	8	10	12
2	4	6	8	10	12	14
4	6	8	10	12	14	16
6	8	10	12	14	16	18
8	10	12	14	16	18	20
10	12	14	16	18	20	22
12	14	16	18	20	22	24

b. What is the probability of obtaining a sum of 14? $\dfrac{1}{6}$ **(8)**

c. What is the probability of obtaining a sum of 24? $\dfrac{1}{36}$ **(8)**

d. What is the probability of obtaining a sum of 11? _0_ **(8)**

e. What is the probability of obtaining an even sum? _1_**(8)**

f. What is the probability of obtaining a sum of 16 and at least one
   cube shows a 6? $\dfrac{1}{18}$ **(8)**

3. If you were designing a fair game using eight-sided solids, what would be your directions? *(6)*

    The directions should contain the idea that there must be an equal chance to win on any throw of the solids. If two eight - sides solids were numbered 1 through 8, and winning is determined by whether a person threw an even or odd sum, then the game would be fair.

4. There are 23 students in Mr. Jones' physical education class. 10 of the students are boys and 13 are girls.
    If Mr Jones randomly chooses one person to set up the activity for the day, what is the probability that:

    a. the person will be a girl? $\dfrac{13}{23}$ *(4)*

    b. the person will be a boy? $\dfrac{10}{23}$ *(4)*

    c. the person will not be a girl? $\dfrac{10}{23}$ *(4)*

    d. Using the idea of complement, explain the relationship of your answer to a to your answer to c. *(6)*

    The sum of the probability of an event and its complement is one. The chance that a girl is not chosen is the complement of the chance that a girl is chosen. The sum of those two probabilities is equal to one.

5. The electricity is out, and Mary needs to get up in the dark and dress for work. She has a green, a blue and a white blouse; a black and a red skirt; a pink, a white and a tan sweater. If Mary chooses a blouse, skirt and sweater randomly, what is the probability that she will be wearing a white blouse, a black skirt and a pink sweater? *(5)*

    The number of combinations of blouses, skirts and sweaters is
    $$3 \cdot 2 \cdot 3 = 18$$

    P(white blouse, black skirt, pink sweater) = $\dfrac{1}{18}$

6. Students in the Connections class at Central High School filled out a survey on their favorite bands. 12 students selected Aerosmith, 15 selected Bon Jovi and 8 selected both. 6 students chose other bands. *(3 each)*

   a. Draw a Venn diagram to illustrate this information.

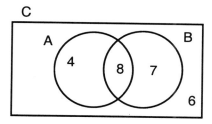

   b. How many students are in the class? _25_

   c. What is the probability that a student selected at random selected Bon Jovi or Aerosmith, but not both? $\dfrac{11}{25}$

   d. What is the probability that a student did not select at random Bon Jovi only? $\dfrac{18}{25}$

   e. Explain how the concept of complement could have been used to solve question d.

   If a student did not select other bands and did not select Aerosmith, that would mean the student selected Bon Jovi only. There are 7 students who selected Bon Jovi only, as shown in the Venn Diagram. These students represent the complement of the question in part d.

MATH *Connections* I
What are the Chances?   Project
Sections 8.4-8.5

The delicatessen at the Towne Centre has two deli clerks to serve customers. The number of customers arriving at the counter varies between 0 and 2 per minute. Customers take a ticket and are served one at a time by the next available deli clerk. It takes each deli clerk three minutes to wait on each customer.

Design a simulation for one 30-minute period.

Directions:
1. Use a random number table or a graphing calculator to generate the number of people arriving each minute by using only the digits 0, 1, and 2 in the table.

2. Conduct the simulation by determining the following for each customer:
    • Specific customer (customer 1, customer 2, etc.)
    • Time at which customer arrives
    • Length of time customer waits in line
    • Time when customer is waited on
    • Time when customer leaves
    NOTE: ALL TIMES REFER TO THE BEGINNING OF THE MINUTE INTERVAL.

3. From the data, develop a table that will include the following:
    • minutes
    • number of customers that arrive at the beginning of each minute
    • number of people waiting in line at the beginning of each minute
    • wait time to be served by the last person in line at the beginning of each minute

Use the information in your table to answer the following questions:
1. What is the length of the waiting time at the beginning of the 10th interval?

2.   How many people are in line at the 15th interval?

3.   What is the average number of people waiting in line over the 30-minute period?

4.   What do you consider a reasonable waiting time for a customer at the deli counter?

5.   You are the deli manager. You are interested in hiring the correct number of deli clerks to service your customers with a minimum of waiting time. Redo the simulation with 3 clerks with the same random numbers which were generated. With this information determine the number of deli clerks needed to run the most efficient and economical deli counter. Explain the reason for your decision.

SIMULATION EXAMPLE

MATH *Connections* I
What are the Chances? Project
Sections 8.4-8.5

0 TO 2 CUSTOMERS
2 CLERKS
3 MINUTE SERVING TIME

Customers	Time of Arrival	Length of Wait	Time Waited on	Time Customer Left
1	1	0	1 (1)	4
2	1	0	1 (2)	4
3	2	2	4	7
4	2	2	4	7
5	4	3	7	10
6	5	2	7	10
7	6	4	10	13
8	6	4	10	13
9	7	6	13	16
10	8	5	13	16
11	8	8	16	19
12	9	7	16	19
13	10	9	19	22
14	11	8	19	22
15	13	9	22	25
16	13	9	22	25
17	14	11	25	28
18	15	10	25	28
19	15	13	28	31
20	18	10	28	31
21	18	13	31	34
22	9	15	31	34
23	21	13	34	37
24	23	11	34	37
25	23	14	37	40
26	24	13	37	40
27	25	15	40	43
28	26	14	40	43
29	27	16	43	46
30	30	13	43	46
31	30	16	46	49

3

SIMULATION EXAMPLE

MATH *Connections* I
What are the Chances? Project
Sections 8.4-8.5

0 TO 2 CUSTOMERS
2 CLERKS
3 MINUTE SERVING TIME

Minutes	Number Arriving	Number Waiting in Line	Waiting Time For Last Person
1	2	0	0
2	2	2	2
3	0	2	1
4	1	1	3
5	1	2	2
6	2	4	4
7	1	3	6
8	2	5	8
9	1	6	7
10	1	5	9
11	1	6	8
12	0	6	7
13	2	6	9
14	1	7	11
15	2	9	13
16	0	7	12
17	0	7	11
18	2	9	13
19	1	8	15
20	0	8	14
21	1	9	13
22	0	7	12
23	2	9	14
24	1	10	13
25	1	9	15
26	1	10	14
27	1	11	16
28	0	9	15
29	0	9	14
30	2	11	16

4

SIMULATION EXAMPLE

MATH *Connections* I
What are the Chances? Project
Sections 8.4-8.5

0 TO 2 CUSTOMERS
3 CLERKS
3 MINUTE SERVING TIME

Customers	Time of Arrival	Length of Wait	Time Waited on	Time Customer Left
1	1	0	1 (1)	4
2	1	0	1 (2)	4
3	2	0	2 (3)	5
4	2	4	4 (1)	7
5	4	0	4 (2)	7
6	5	0	5 (3)	8
7	6	1	7 (1)	10
8	6	1	7 (2)	10
9	7	1	8 (3)	11
10	8	2	10 (1)	13
11	8	2	10 (2)	13
12	9	2	11 (3)	14
13	10	3	13 (1)	16
14	11	2	13 (2)	16
15	13	1	14 (3)	17
16	13	3	16 (1)	19
17	14	2	16 (2)	19
18	15	2	17 (3)	20
19	15	4	19 (1)	22
20	18	1	19 (2)	22
21	18	2	20 (3)	23
22	19	3	22 (1)	25
23	21	1	22 (2)	25
24	23	0	23 (3)	26
25	23	2	25 (1)	28
26	24	1	25 (2)	28
27	25	1	26 (3)	29
28	26	2	28 (1)	31
29	27	1	28 (2)	31
30	30	0	30 (3)	33
31	30	1	31 (1)	34

SIMULATION EXAMPLE

MATH *Connections* I
What are the Chances? Project
Sections 8.4-8.5

0 TO 2 CUSTOMERS
3 CLERKS
3 MINUTE SERVING TIME

Minutes	Number Arriving	Number Waiting in Line	Waiting Time For Last Person
1	2	0	0
2	2	1	2
3	0	1	1
4	1	0	0
5	1	0	0
6	2	2	1
7	1	1	1
8	2	2	2
9	1	3	2
10	1	2	3
11	1	2	2
12	0	2	1
13	2	2	3
14	1	2	2
15	2	4	4
16	0	2	3
17	0	1	2
18	2	3	2
19	1	2	3
20	0	1	2
21	1	2	1
22	0	0	0
23	2	1	2
24	1	2	1
25	1	1	1
26	1	1	2
27	1	2	1
28	0	0	0
29	0	0	0
30	2	1	1

SOLUTION KEY AND SCORING GUIDE

MATH *Connections* **I**
WHAT ARE THE CHANCES? QUIZ
SECTIONS 8.4 - 8.5
PROJECT

The simulation may be developed from a random number table or by using the TI-82. The number of customers each minute was generated for the sample simulations on the following pages by using a random number table. The numbers were determined by selecting a number randomly as a starting point then using only 0's, 1's and 2's as they appear in order from the table. The numbers were:

    2 2 0 1 1 2 1 2 1 1
    1 0 2 1 2 0 0 2 1 0
    1 0 2 1 1 1 1 0 0 2

The delicatessen at the Towne Centre has two deli clerks to serve customers. The number of customers arriving at the counter varies between 0 and 2 per minute. Customers take a ticket and are served one at a time by the next available deli clerk. Each deli clerk services one customer in three minutes.

Design a simulation for one 30-minute period.

Directions:
1. Use a random number table or a TI-82 to generate the number of people arriving each minute by using only the digits 0, 1, and 2 in the table.

2. Conduct the simulation by determining the following for each customer:
    • Specific customer (customer 1, customer 2, etc.)
    • Time at which customer arrives
    • Length of time customer waits in line
    • Time customer is served
    • Time customer leaves
    NOTE: ALL TIMES REFER TO THE BEGINNING OF THE MINUTE INTERVAL.

3. From the data, develop a table that will include the following:
    *   minutes
    *   number of customers that arrive at the beginning of each minute
    *   number of people waiting in line at the beginning of each minute
    *   wait time to be served by the last person in line at the beginning of each minute

Scoring: Give 50 points for the simulation, based upon organization of the information, use of appropriate tables, clarity of information and usefulness of results. The other 50 points will be distributed as follows. Answers to the questions below will be consistent with the specific simulation completed by the students.

Use the information in your table to answer the following questions:
1. What is the length of the waiting time at the beginning of the 10-minute interval? *(7)*

2. How many people are in line at the 15-minute interval? *(7)*

3. What is the average number of people waiting in line over the 30-minute period? *(7)*

4. What do you consider a reasonable waiting time for a customer at the deli counter? *(7)*

5. You are the deli manager. You are interested in hiring the correct number of deli clerks to service your customers with a minimum of waiting time. Redo the simulation with 3 clerks with the same random numbers which were generated.   With this information determine the number of deli clerks needed to run the most efficient and economical deli counter.   Explain the reason for your decision. *(22)*

MATH *Connections* I
What are the Chances?   Test
Chapter 8 (A)

Name _____ Date _____

1. You will be given a two sided object (such as a coin or 2-colored disk).
   You will "toss" the object 10 times.

   a. Predict the number of heads (or a color) that you will get. _____

   b. "Toss" the object 10 times and keep a tally of the outcomes. List the actual
      outcomes. (For example: H H T H T T T H H H)

   c. Explain the relationship between your actual results and your prediction.

   d. Write your results on the board (or on the overhead) as directed by your teacher
      after all the results of the class have been recorded , write out your explanation,
      telling how your outcomes compare to the total class outcomes for the
      experiment.

2. For each of the following events, select a number on the line below the statement which indicates the probability you would assign to the event (E). Write your answer in "P(E) = *number*" form and explain your choice.

    a. I will swim in the ocean this summer.

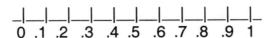

    Explain your choice:

    b. I will drive in a Jaguar before I am 18.

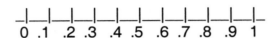

    Explain your choice:

    c. It will eat lobster on my birthday.

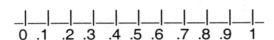

    Explain your choice:

3. Identify an event or situation that might have a P(E) = .8.

4. The numbers 1 to 50 are written on separate slips of paper and placed in a bowl. After the slips are thoroughly mixed, one of the slips is selected from the bowl without looking into it.

     a. What is the probability that the number on the slip is 25? _____

     b. What is the probability that the number on the slip is a multiple of 5?
       _____

     c. What is the probability that the number on the slip is a multiple of 5 and a multiple of 3? _____

     d. What is the probability that the number on the slip is greater than 12?
       _____

     e. What is the probability that the number on the slip is divisible by 9?
       _____

5. Each of the nine letters in the word "Tennessee" are written on separate slips of paper and placed in a bowl. The slips of paper are thoroughly mixed and one of them is picked at random.

     a. What is the probability that an "S" is picked? _____

     b. What is the probability that a vowel is picked? _____

     c. What is the probability that a "Q" is picked? _____

     d. What is the probability that the letter picked is in the alphabet?
       _____

6. The students at Central High School are to be given identification ID numbers. The ID numbers are to contain 4 digits; zero cannot be the first number. What is the probability of the ID number being 1111? _____

SOLUTION KEY AND SCORING GUIDE

MATH *Connections* I
What are the Chances? Test
Chapter 8 (A)

1. You will be given a two sided object (such as a coin or 2-colored disk). You will "toss" the object 10 times. *(20 points)*

   a. Predict the number of heads (or a color) that you will get.

      It is expected that the student will predict that the number of heads will be 5 since the probability of tossing a head is one-half.

   b. "Toss" the object 10 times and keep a tally of the outcomes. List the actual outcomes. (For example: H H T H T T T H H H)

      The results will vary since it is an experiment.

   c. Explain the relation between your actual results and your prediction.

      The answers will be dependent on the prediction and the actual results of the experiment.

   d. Write your results on the board (or on the overhead) as directed by your teacher. After all the results of the class have been recorded , write out your explanation, telling how your outcomes compare to the total class outcomes for the experiment.

      It is expected that the results of many events will produce an outcome which will demonstrate that the theoretical probability of one-half will be realized. In other words, one-half of the number of tosses should be a heads. The student should allude to this in his/her response.

2. For each of the following events, select a number on the line below the statement which indicates the probability you would assign to the event (E). Write your answer in "P(E) = *number"* form and explain your choice. *(5 each)*

> *It is expected that the student will demonstrate an understanding how numbers are assigned to probable events.*

a. I will swim in the ocean this summer.

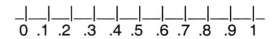

Explain your choice:

b. I will drive in a Jaguar before I am 18.

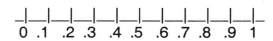

Explain your choice:

c. It will eat lobster on my birthday.

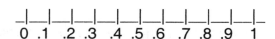

Explain your choice:

3. Identify an event or situation that might have a P(E) = .8. *(6)*

> *It is expected that the student will discuss what is meant by a probability of 0.8. Answers will vary as most of the examples used to this point have not had an easily obtainable probability of 0.8.*

4. The numbers 1 to 50 are written on separate slips of paper and placed in a bowl. After the slips are thoroughly mixed, one of the slips is selected from the bowl without looking into it. *(6 each)*

    a. What is the probability that the number on the slip is 25? $\dfrac{1}{50}$ *or 0.02*

    b. What is the probability that the number on the slip is a multiple of 5? $\dfrac{10}{50}$ *or 0.2*

    c. What is the probability that the number on the slip is a multiple of 5 and a multiple of 3? $\dfrac{3}{50}$ *or 0.06*

    d. What is the probability that the number on the slip is greater than 12? $\dfrac{38}{50}$ *or 0.76*

    e. What is the probability that the number on the slip is divisible by 9? $\dfrac{5}{50}$ *or 0.1*

5. Each of the nine letters in the word "Tennessee" are written on separate slips of paper and placed in a bowl. The slips of paper are thoroughly mixed and one of them is picked at random. *(6 each)*

    a. What is the probability that an "S" is picked? $\dfrac{2}{9}$ *or 0.22*

    b. What is the probability that a vowel is picked? $\dfrac{4}{9}$ *or 0.44*

    c. What is the probability that a "Q" is picked? 0

    d. What is the probability that the letter picked is in the alphabet? 1

6. The students at Central High School are to be given identification numbers (ID). The ID numbers are to contain 4 digits; zero cannot be the first number. What is the probability of the ID number being 1111? _____ *(5)*

    The possible combinations are 9 • 10 • 10 • 10 or 9000.

    P(1111) = $\dfrac{1}{9000}$